Advanced Coatings for the Corrosion Protection of Metals

Diana Petronela BURDUHOS-NERGIS

Dumitru-Doru BURDUHOS-NERGIS

Simona-Madalina BALTATU

Petrica VIZUREANU

Gheorghe Asachi Technical University of Iasi, Faculty of Materials Science & Engineering, Romania

Published by **Materials Research Forum LLC**
Millersville, PA 17551, USA

Published as part of the book series
Materials Research Foundations
Volume 115 (2022)
ISSN 2471-8890 (Print)
ISSN 2471-8904 (Online)

Print ISBN 978-1-64490-166-3
ePDF ISBN 978-1-64490-167-0

Distributed worldwide by

Materials Research Forum LLC
105 Springdale Lane
Millersville, PA 17551
USA
http://www.mrforum.com

Printed in the United States of America
10 9 8 7 6 5 4 3 2 1

Table of Contents

Introduction 7

Metals Corrosion 1
   1.1. Metal corrosion 1
   1.2. Corrosion processes classification 2
      1.2.1. Chemical corrosion 3
      1.2.2. Electrochemical corrosion 8
      1.2.3. Biochemical corrosion 12
      1.2.4. Continuous corrosion 13
      1.2.5. Discontinuous corrosion 14
      1.2.6. Intercrystalline corrosion 17
      1.2.7. Transcrystalline corrosion...17
      1.2.8. Selective corrosion..18
      1.2.9. Atmospheric corrosion..18
      1.2.10. Underground corrosion 19
      1.2.11. Aqueous corrosion 19
      1.2.12. Corrosion due to mechanical stress.................................20
      1.2.13. Microbiological corrosion...21
   1.3. Metal passivation 21
   References 23

Methods for Testing and Assessing Corrosion Behaviour 29
   2.1. Quantitative (direct) analysis of the corrosion behaviour of
   materials 29
   2.2. Indirect analysis of the corrosion behaviour of materials.....................31
      2.2.1. Corrosion assessment by non-destructive methods 32
      2.2.2. Corrosion assessment by electrochemical methods.....................36
   References 50

Methods of Corrosion Protection by Depositing Metallic Protective
Layers 57
   3.1. Introduction 57

3.2. Preparation of metal surfaces for coating 58
3.3. Electrolytic deposition of protective layers 61
 3.3.1. Zinc plating 64
 3.3.2. Tin plating 68
 3.3.3. Chrome plating 73
 3.3.4. Nickel plating 78
 3.3.5. Copper plating 81
 3.3.6. Silver plating 84
 3.3.7. Gold plating 85
 3.3.8. Cadmium plating 86
 3.3.9. Lead plating 87
 3.3.10. Aluminium plating 87
 3.3.11. Iron plating 87
 3.3.12. Brass plating 88
3.4. Spray metal coatings 89
3.5. Diffusion coatings 90
3.6. Cladding coatings 91
References 92

Methods of Corrosion Protection by Depositing Inorganic and Organic Protective Layers 103
4.1. Introduction - inorganic protective layers 103
4.2. Phosphating 103
4.3. Oxidation 108
 4.3.1. Oxidation of ferrous metals (Blackening/Bluing) 109
 4.3.2. Chromating 110
 4.3.3. Oxidation of aluminium (anodizing) 111
4.4. Enamelling 112
4.5. Introduction –organic protective layers 113
4.6. Painting or varnishing 114
4.7. Bituminous coatings 116
References 116

Advanced Methods for Deposition of Protective Thin Layers .. 121

  5.1   Introduction 121

  5.2   PVD method 121

       5.2.1. Deposition of layers by thermal evaporation 126

       5.2.2. Cathodic spray deposition 128

       5.2.3. Deposition of layers by ion plating 129

  5.3. CVD method 131

  5.4   Deposition of anticorrosive thin layers 133

  5.5   Deposition of hard, lubricating and wear-resistant thin layers 134

  5.6   Decorative thin film deposition 135

  References 135

About the Authors 142

Introduction

The corrosion resistance of metals is an important issue in many industrial fields. This leads to high economic losses, so adequate protection against corrosion is important in many applications. Therefore, coatings are the most important technology for protection against corrosion of metal surfaces.

In the last years, the field of coatings used to improve the corrosion resistance of the metal surface have gained more importance. Many researchers study the corrosion mechanism and the properties of traditional coatings. Also, due to environmental issues, new coatings are developed and analysed.

The book contains a complex study in the field of materials science and chemistry on improving the corrosion resistance properties of metals. At the same time, it contains advanced notions about the development and characterization of several types of coatings such as organic and inorganic coatings, metallic coatings, pretreatments etc.

Therefore, the book "Advanced Coatings for the Corrosion Protection of Metals" is relevant for fundamental and applied research in the field of materials engineering because it shows the importance of coatings for the corrosion protection of metallic surfaces describing the metal corrosion processes and presenting some methods for testing and assessing the corrosion behaviour of the metals. It also describes how to design and develop different types of coatings, the properties of coating materials and summarize the new developments in this field.

The first chapter, "Metals corrosion", presents introductory notions related to different types of corrosion processes, as well as about the passivation of metals.

In the second chapter, "Methods for testing and assessing corrosion behaviour", are presented different methods used to determine the corrosion behaviour of metals, such as quantitative (direct) analysis and indirect analysis (optical and SEM microscopy, laser profilometry, time of flight diffraction, ellipsometry, linear polarization, cyclic polarization, electrochemical impedance spectroscopy etc.)

The third chapter, "Methods of corrosion protection by depositing metallic protective layers", describes different types of metallic coatings, such as electrolytic coatings, spray metal coatings, diffusion coatings and cladding coatings.

The fourth chapter, "Methods of corrosion protection by depositing inorganic and organic protective layers", contains information about the different methods to deposit inorganic layers on the metal surface such as phosphating, oxidation and enamelling. It also presents information about organic protective layers deposited by painting or varnishing, as well as bituminous coatings.

The last chapter, "Advanced methods for deposition of protective thin layers", presents the different methods used to obtain thin layers with high properties. It also contains information about the anticorrosive thin layers, decorative thin layers, hard, lubricating and wear-resistant thin layers

Chapter 1

Metals Corrosion

1.1. Metal corrosion

Corrosion is the spontaneous phenomenon of partial or total destruction of materials, especially metals, following chemical, electrochemical or biochemical reactions to the interaction with the environment [1]. The technical-economic aspect of corrosion results from the fact that approximately 15-18% of the amount of extracted and processed metal is degraded annually.

Depending on the nature and structure of the metallic phase, the chemical composition of the corrosive environment and the conditions of the corrosion process, the reaction mechanism and the intensity of degradation of the metallic material are different [2].

Many factors influence the corrosive environment, such as humidity, pH, oxygen concentration and accessibility, the presence of ions with a specific action, the concentration of the metal ion, the electroconductive properties of the environment, the presence of inhibitors etc. [3].

From a thermodynamic point of view, the tendency of metals to corrode is characterized by the free enthalpy (G) of the process, its variation providing data on the possibility or impossibility of the corrosion process [4].

The most noticeable effects of corrosion are weight change, surface alteration and decrease in the mechanical properties of the material.

The negative economic consequences are given by the cost of corrosion, which includes direct losses related to the replacement of parts and equipment and indirect losses related to the stagnation of production by stopping machinery and installations. Losses of corrosive metal materials account for 1/3 of world production, so researchers in the field are concerned with finding solutions either by developing materials with anticorrosive properties or by protecting existing materials with various methods of protection against corrosion (coatings). By preventing corrosion, it has been found that about 25% of annual corrosion-related expenses can be eliminated [5].

The reduction of costs can be done either in the design phase of the equipment by choosing materials with high corrosion resistance properties or by using coatings that provide these properties to the material.

To reduce corrosion, anti-corrosion protection measures are taken by artificially creating kinetic obstacles in the way of the electrode process of the corrosion system by alloying or cathodic polarization of the installation. There is no universal method of corrosion protection.

The choice of method is made in the design, depending on the properties of the corrosive environment, the type of construction, functional requirements and the working time of the machine.

1.2. Corrosion processes classification

In general, metals and metal alloys are subject to the effect of corrosion degradation, but degradation can also occur on plastics through physical or chemical ageing, as well as the degradation of ceramic materials (concrete and limestone).

Corrosion of metallic materials is due to their thermodynamic instability in relation to their oxidized form [6], so it is a natural phenomenon by which metals tend to return to a more stable state, lower energy of oxides, chlorides, sulfates etc. The classification of corrosion processes is done according to certain criteria (Fig. 1.1).

The criteria for classifying corrosion processes are as follows:

a) according to the deployment mechanism: chemical corrosion [7], electrochemical corrosion [8] and biochemical corrosion [9];

b) according to the attack distribution: general corrosion (continuous) [10], when the entire surface was covered by the action of the corrosive environment and localized corrosion (discontinuous) [11] when corrosion occurs on certain parts of the material;

c) according to the type of corrosive environment: wet corrosion and dry corrosion [12];

d) according to the character of the destruction in relation to the material structure: intercrystalline corrosion, transcrystalline corrosion and selective corrosion [13];

e) depending on the specific conditions or according to the place where it appears: atmospheric corrosion [14], underground corrosion [15]; aqueous corrosion (in water) [16], microbiological corrosion [17] and corrosion due to mechanical stress [18].

The classification of corrosion in chemical and electrochemical processes is not absolute. Chemical corrosion is transformed into electrochemical corrosion by the condensation of water vapour.

The corrosion type, as well as the intensity of material destruction, are determined by the nature and structure of the metal phase, by the chemical composition of the corrosive environment and by the conditions of the technological process.

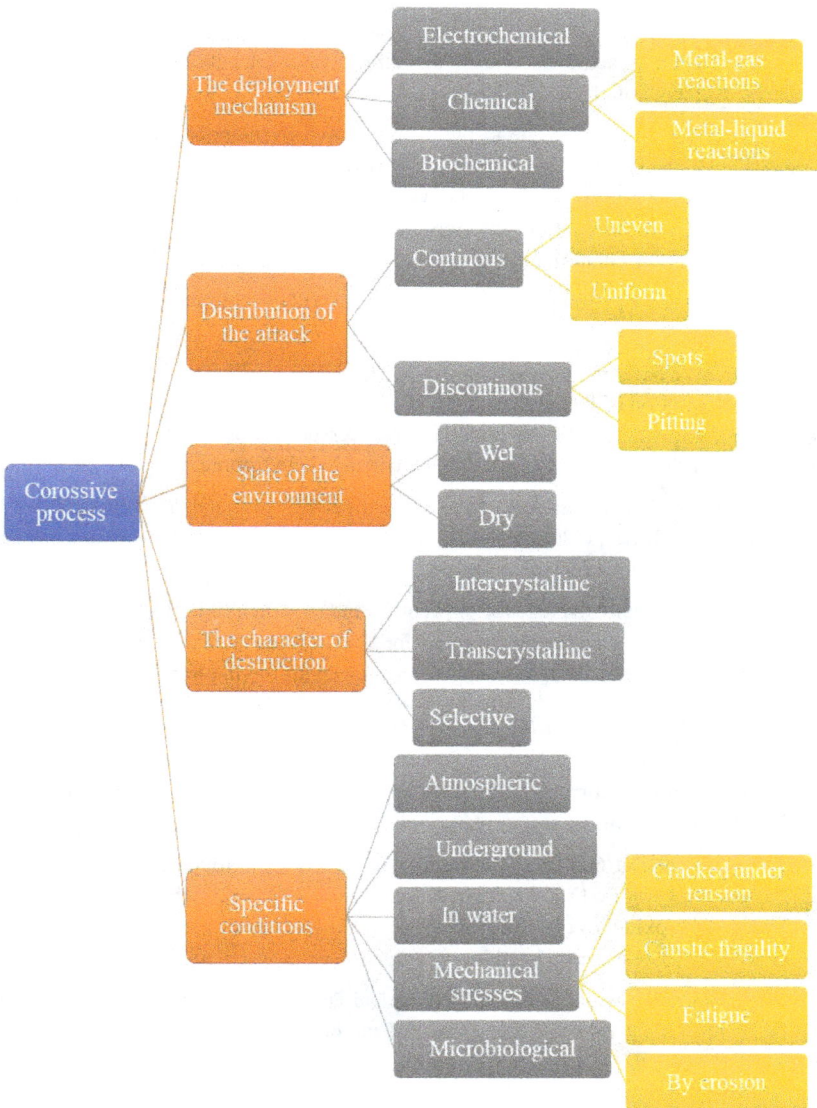

Figure 1.1.   Classification of corrosion processes.

1.2.1. Chemical corrosion

Chemical corrosion is achieved due to the direct chemical action of the environment without the exchange of electrical charges.

Chemical corrosion can occur due to the affinity between materials and some dry gases (O_2, SO_2, CO, CO_2, H_2 etc.) or with non-electrolyte media (gasoline, alcohols, benzene, etc.). The chemical reaction that forms between the material and the gas or liquid leads to material losses, the disintegration (disaggregation) of the material by the salt crystals that form in the pores of the material [19].

The intensity of the corrosion process depends on several factors, such as:

- the material type;
- the corrosive environment type;
- the concentration of the environment;
- the environmental conditions (temperature, pressure);
- the contact time of the material with the corrosive environment.

Typical examples of chemical corrosion are the processes of metals oxidation, as well as the formation of sulfides, chlorides, iodines etc. if the corrosive agent contains chemical elements such as sulfur, chlorine, iodine etc.

Through the external corrosive agents, oxygen has the most harmful effect on metals. If the clean surface of most metals is exposed to air for a while, the material oxidizes rapidly if the oxidation reaction between metal and oxygen occurs with decreasing free energy (eq. 1.1).

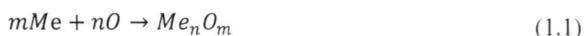

$$mMe + nO \rightarrow Me_nO_m \qquad (1.1)$$

In the case of chemical corrosion, proportional with the increase of the temperature, the corrosion rate also increases.

For example, even if at 600 °C a complex layer of oxides is found on the steel surface (cinder - FeO, Fe_2O_4 și Fe_3O_4), when hot deformation of the steel, where the temperatures exceed 900 °C, an oxide layer is formed on the metal surface, of about 1-2 mm leading to significant metal losses at each heating.

In a dry atmosphere, corrosion is purely chemical and is limited to the formation of the oxide layer that affects only the appearance of the outer surface of the material, without strongly influencing the mechanical properties.

Chemical corrosion thermodynamic

The spontaneity of the corrosion reaction (oxidation) is expressed by the negative variation of the free enthalpy (G), in isobaric-isothermal conditions (constant pressure and temperature). The value of the free enthalpy of reaction characterizes the affinity of metals to oxygen and corresponds to the oxide formation reaction, starting from the pure metal and oxygen [20].

Depending on the value of the free enthalpy of the oxidation reaction of the most used metals, the following sequence of decreasing the oxidation tendency can be established:

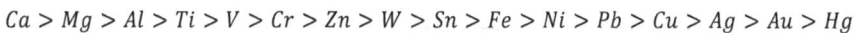

$$Ca > Mg > Al > Ti > V > Cr > Zn > W > Sn > Fe > Ni > Pb > Cu > Ag > Au > Hg$$

The direction of this reaction depends on the magnitude of the partial pressure of oxygen and the dissociation voltage of the metal oxide at the given temperature. If the partial pressure of oxygen is higher than the dissociation voltage of the oxide, corrosion of the metal is thermodynamically possible.

On the contrary, if the dissociation voltage of the oxide is higher than the partial pressure of oxygen, the oxide will no longer be stable under these conditions.

At equilibrium, the two pressures will be equal. The partial pressure of oxygen in the air, at atmospheric pressure, is considered constant and equal to 0.2 at. It can be expressed that in the clean atmosphere are thermodynamically possible those oxidation reactions that lead to the formation of oxides with dissociation stress less than 0.2 at. [20].

In order to reduce the affinity of the metal for oxygen and to reduce the oxide, partial pressure of oxygen lower than the dissociation voltage of the oxide at room temperature must be ensured. Such conditions are mentioned in order to attenuate the corrosion to the heat treatment of some alloys, under the conditions of using a protective environment.

The formation and growth of surface oxide films are governed by the laws of crystallo-chemical reactions and the phenomena of electrical conductivity and mass transfer (substance).

For the oxide to form, both the ionization of the metal atom in the material and the oxygen molecule must take place. The ionization of the metal most likely occurs at the metal-oxide interface, it can be easily assimilated to the anode of a galvanic element, and the oxide-air interface (oxygen), where the reduction reaction or similar, takes place can be assimilated to the cathode of the galvanic element [21].

The oxide layer formed between the two electrodes acts as an electrolyte, as can be seen in Fig. 1.2.

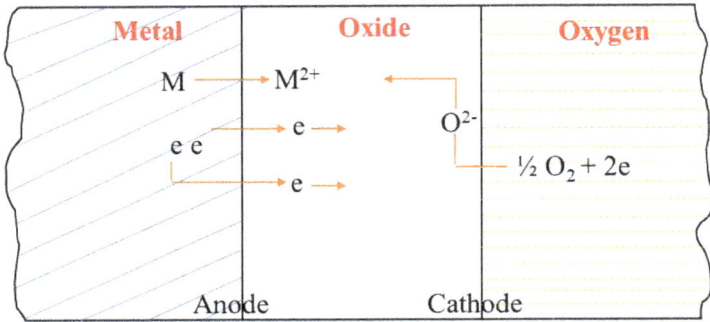

Figure 1.2.   The metal ionization and the oxygen molecule

The crystalline-ionic structure of the oxide film is characterized by a certain conductivity, which depends on both the temperature and the stoichiometric composition [22].

There are films characterized by ionic conductivity, determined by either cations or anions. Also, are films whose conductivity is due to both anions and cations.

Chemical corrosion kinetics

The speed of the corrosion process is given by the diffusion of cations and anions through the film and depends on the imperfections of the crystal lattice of the metal and the oxides or compounds in the film [23].

Depending on the growth rate of the films formed during oxidation in the atmosphere, metals can oxidize according to a linear, parabolic or logarithmic law [20].

a) Metals that cannot form protective films, whether they are not continuous, such as alkali metals, or oxidation products are volatile, as is the case at high temperatures with vanadium, molybdenum, tungsten, osmium and iridium, oxidizes according to a linear law, with constant speed over time (eq. 1.2):

$$\frac{dy}{dt} = k \qquad (1.2)$$

where: y- film thickness;

$\quad$ t- oxidation time;

$\quad$ k -constant.

Integrating the equation, we obtain the linear law of corrosion (eq. 1.3):

$$y = k \cdot t + c \qquad (1.3)$$

If it is considered that oxidation begins on a clean surface of the metal, the integration constant $c = 0$, and $y = k \cdot t$.

According to the linear law, the oxidation of the metals of this group would take place at the speed of the chemical reaction between metal and oxygen, the film not being able to break the corrosion.

However, in some cases, although over time the growth of the films is observed in proportion to the time, it is found that the effective corrosion rate is much lower than the speed of the direct reaction between metal and oxygen. According to the same law, alloys with a high content of metals in this category are also oxidized [20,24].

b) In the case of many metals, in which the chemical oxidation produces continuous films, the corrosion attenuates with time, with the increase of the film thickness. Experimentally, it has been established that iron, between 500-1100 °C, copper between 300-1000 °C, nickel in the atmosphere, silver in vapours and iodine solution corrode at a speed inversely proportional to the thickness of the film. In these cases, corrosion is subject to parabolic law (eq. 1.4):

$$\frac{dy}{dt} = \frac{k'}{y} \qquad (1.4)$$

which becomes integrated: $y^2 = 2k' \cdot t + c'$ or $y^2 = k \cdot t$.

where y - film thickness;

  t - oxidation time

  k – constant;

  c' - constant of integration, which in the case of the initial non-corroded metal surface is equal to 0.

The parabolic law is observed for the chemical corrosion of metals at high temperatures [24].

There are deviations from the parabolic law of oxidation that can occur either by partial destruction on the surface of thicker corrosion films due to accumulated internal stresses or due to the occurrence of additional causes of diffusion braking (for example, lowering the diffusion coefficient of reactants with oxide layer thickness) [22,25].

Metals such as cobalt, manganese, beryllium, zinc and titanium belong to this category.

Aluminium, chromium, silicon and zinc at various temperatures, as well as many other metals at low temperatures oxidize much more slowly than the metals of the previous groups, being very effectively protected from very fine corrosion films. The oxidation of these metals is subject to a logarithmic law.

Thus, the oxidation rate of Al, Cr, Si and Zn, of iron up to 375 °C and nickel up to 650 °C, varies according to the expression (eq. 1.5):

$$\frac{dy}{dt} = \frac{k}{e^y} \qquad (1.5)$$

or integrated: $y = \ln kt$.

Intense corrosion braking, even at small thicknesses of the corrosion product layer, is considered to be due to additional causes such as:

- decrease the electrical conductivity of the film;
- reducing the diffusion through the film;
- compaction of the film due to strong internal compression efforts.
- The process of chemical corrosion is influenced by many factors, which fall into two categories:
- External factors, of which the most important are: temperature, pressure, the speed of the corrosive agent, corrosive environment chemical composition, etc.
- Internal factors related to the nature, chemical composition and structure of the material, surface condition, etc.

One of the main characteristics of chemical corrosion is that the corrosion products generally remain at the site of the interaction of the metal with the corrosive agent in the form of films of different thicknesses [26].

For the film of corrosion products to be able to protect the metal from corrosion, it must have the following properties: to have a medium thickness, to adhere to the metal, to be compact and continuous, waterproof, elastic, resistant and to have a coefficient of expansion equal to that of the metal, so that at temperature variations it does not form cracks. Thin films generally cannot protect the metal to a sufficient extent due to low mechanical strength, and thick films crack easily due to internal stresses [27,28].

1.2.2. Electrochemical corrosion

Electrochemical corrosion is achieved by the reaction between a metal and an electrolyte with which it is in contact. This is the destructive attack exercised on a metal or alloy by the corrosive environment (electrolytes in solution or melt, water, vapour, liquid metal, etc.) by means of an electrochemical reaction, which involves a transfer of electrons and ions to the metal interface (electronic conductor) - electrolyte (ionic conductor), under the influence of an electric potential (of electrode) existing between phases [29].

For the occurrence of electrochemical corrosion it is necessary to have an anode, a cathode, an electrolyte and a conductor, so a galvanic element.

By introducing the material into water or an environment with electrolytic properties, galvanic elements appear on the metal surface, in which metal impurities function as micro-cathodes, with hydrogen discharge on their surface, while the material, acting as an anode, dissolves.

Typical examples of electrochemical corrosion are found in steel objects exposed to atmospheric corrosion (iron rust), as well as corrosion caused by electric scattering currents in the soil, also called stray currents (tram rails or electric rail).

Basically in the case of corrosion, it is important to know the corrosion rate. If the corrosion rate is very low, even if corrosion occurs, the material can be considered resistant to corrosion.

The corrosion rate is expressed by the mass of material destroyed per unit area per unit time (g / m^3h) or the depth reached by the degradations per unit time (mm / year) [30].

Knowing these indices allows the appropriate choice of material according to the nature of the corrosive environment and the sizing of installations according to the duration of the operation.

Corroded metal is generally in the form of metal constructions and installations, the cost of which is much higher than the cost of the material from which they are made. Along with the cost of the material, very large amounts are spent on:

- repair of defects caused by corrosion;
- oversizing to prevent accidents;
- use of expensive anticorrosive materials (non-ferrous metals and alloys, special steels);
- application of current methods of corrosion protection etc.

In most cases, the corrosion encountered in the industry is electrochemical. An electrochemical corrosion system metal-corrosive solution is characterized by the existence of an electric current generated by electrochemical processes that take place at the limit of metal-solution separation.

The underlying cause of all electrochemical corrosion processes is the tendency of most metals to pass into solution in the form of ions when they come in contact with an electrolyte, while the mechanism of local currents must be seen as one of the most important possible routes. practice in the development and acceleration of corrosion.

Electrochemical corrosion processes take place on the surface of the metal in the form of two electrode reactions: an anodic process and a cathodic one. These reactions take place simultaneously with the participation of an equal number of electrons and are independent of each other, respectively one process cannot influence through this, the kinetics of the other process.

The electrochemical process of corrosion occurs by carrying out two types of simultaneous parallel reactions: an anodic ionization reaction (oxidation or corrosion of the metal itself) and a cathodic reaction to reduce an agent capable of accepting electrons remaining in the metal phase by passing ions in solution.

The cathodic process can be:

- In acidic environment - when hydrogen ions are reduced (corrosion with hydrogen depolarization)
- In alkaline environment - when the reduction of dissolved oxygen in the electrolyte is achieved (corrosion with oxygen depolarization)
- In the neutral environment - when there is a reduction of an oxidant (corrosion with depolarization of oxidizing metal ions).

Electrochemical corrosion thermodynamics

Depending on the negative or positive value of the free reaction enthalpy we can analyze the possibility of oxidation of the material (for example, if the free reaction enthalpy is negative, it indicates the possibility of oxidation of the material, and if it is positive, the oxidation will not be possible thermodynamically) [22].

From an electrochemical point of view, the possibility of a metal dissolving in a corrosive solution can be explained by the electromotive force of the galvanic element in which the anodic and cathodic corrosion reaction is reversibly performed. The higher the electromotive force of the galvanic element, the greater the possibility of corrosion reactions [20]. The electromotive force of a galvanic element represents the relationship between the two electrode potentials $\varepsilon_c - \varepsilon_a$.

So, the process of electrochemical corrosion is thermodynamically possible if the reversible potential of the metal in the given solution is more electronegative than the reversible potential of an oxidizing agent, from the same solution, which can be reduced. The calculation is therefore limited to determining and comparing the equilibrium potentials ε_c și ε_a. The obtained result allows us to appreciate only the corrosion tendency of the metal because the thermodynamics cannot provide data on the effective speed and the corrosion mechanism.

From a practical point of view, first of all, the effective speed of corrosion is of interest. If the process is possible but takes place very slowly, the metal can be considered resistant to corrosion.

Electrochemical corrosion kinetics

Electrochemical corrosion reactions as electrode processes take place at the boundary of the two phases, metal-solution and comprise several stages:

- the transport of the corrosive agent to the reaction surface;
- the phase limit reaction;
- the transport of reaction products (corrosion) to the solution volume.

The overall speed of the process is determined by the speed of the slowest stage (limiting stage of the process). For example, at the corrosion of metals in acidic environments, the determining speed is the reaction rate of the phase limit [22].

In general, the higher the corrosion rate, the higher the electromotive force, the lower the resistance and the faster the products resulting from the electrochemical reactions of the corrosion are removed by dissolution or release. Stagnation of corrosion products on the metal surface can produce polarization phenomena, which reduces the potential difference and therefore reduces or stops the corrosion of the metal.

Polarization is defined as the difference between the equilibrium (maximum) potential of an electrode and the actual (regime) potential. The practical result of polarization is the intense braking of corrosion, its speed can decrease tens and hundreds of times compared to the initial value. Therefore, due to its special practical importance, the study of polarization is the main object of electrochemical corrosion kinetics [20].

The main types of polarization, encountered in electrochemical corrosion reactions are:

- Concentration (diffusion) polarization;
- Activation polarization (electrochemical, charge transfer, overvoltage).

Diffusion polarization is due to a decrease in the concentration of ions around the cathode following a process of reducing or increasing the concentration of ions around the anode by dissolving it. The result of the concentration changes is materialized by shifting the potential to the negative or positive range, relative to the value of the equilibrium potential.

The activation polarization is caused by the electrode process itself (anodic or cathodic) for which additional energy (an overvoltage) is required.

The value of the activation polarization is of particular importance in the case of electrode reactions involving hydrogen and oxygen. Activation polarization, respectively hydrogen overvoltage is strongly influenced by the nature of the metal, current density, temperature, solution composition, etc.

In general, the process of electrochemical corrosion is influenced by many factors, which can be grouped into two categories: internal factors and external factors - which can act either separately or simultaneously with complex action.

Among the external factors, the most important, we list the pH of the solution, the chemical composition of the aggressive environment, respectively the electrical conductivity, temperatures, agitation of the environment, dispersion currents etc.; and among the internal factors, we mention the nature of the metal and its structure, the condition of the metal surface, the deformations and internal stresses of the metal.

Because the resulting influence of a large number of factors is difficult to predict, in practice the aim is often to establish empirical corrosion-time relationships. Based on experimental data, two types of curves can be made: weight loss (y) – time (t) and corrosion rate $(dy\,/\,dt)$ – time (t) [20].

The properties and behaviour of the corrosion process are of particular importance in terms of the nature and kinetics of the corrosion reaction. Thus, when the corrosion process is soluble it does not have a protective action. At higher pH values, following the dissolution of the metal in the corrosive environment, hardly soluble combinations can be formed, which, accumulated on the metal surface, may or may not protect it, as the deposited film is compact or permeable.

1.2.3. Biochemical corrosion

Biochemical corrosion (biocorrosion) is determined by the metabolic or excretory activity of various microorganisms, it usually accompanies electrochemical corrosion; as a rule, biocorrosion occurs in ferrous metal structures (except in non-ferrous ones) in contact with stagnant water [31].

It is generally believed that corrosion is a process that can only occur when the metal comes in contact with water and oxygen. A significant part of corrosion occurs due to the total lack of oxygen. Black dots - iron (II) sulfide can be seen on the surface of the pipes thus corroded. If iron (II) sulfide is removed, an anodic recess can be seen whose surface is hollow iron.

For biocorrosion, on iron surfaces covered with water and/or biological deposits, sulfate-reducing bacteria are primarily responsible. Such media contain sulfate ions but not oxygen.

Also, for the appearance of biocorrosion, another group of microorganisms that live in the same environment in oxygen-free environments may be responsible. They cover their energy needs by oxidizing hydrogen with carbon dioxide [9]. The result of this reaction is methane and water. These methane-producing bacteria live in significant quantities in oxygen-free environments, such as: under deposits of technical sludge on the bottom of tanks, under deposits in tubes and pipes with slow draining etc.

This type of corrosion is caused by microorganisms that can use the metal as a nutrient medium. The products of the vital activity of microorganisms can also be aggressive.

Biochemical corrosion is determined by the activity of microorganisms that either use the metal as a chemical culture medium or eliminate highly corrosive products due to their metabolism [32].

In general, biocorrosion occurs in ferrous metal structures that are in contact with stagnant water, it is less pronounced in non-ferrous materials.

1.2.4. Continuous corrosion

Continuous corrosion occurs on the entire surface of the metal uniform (when the corrosion has the same depth, Fig. 1.3 (a)) or non-uniform (when due to the different corrosion rate, the corrosion extends to different depths, Fig. 1.3 (b)) [33].

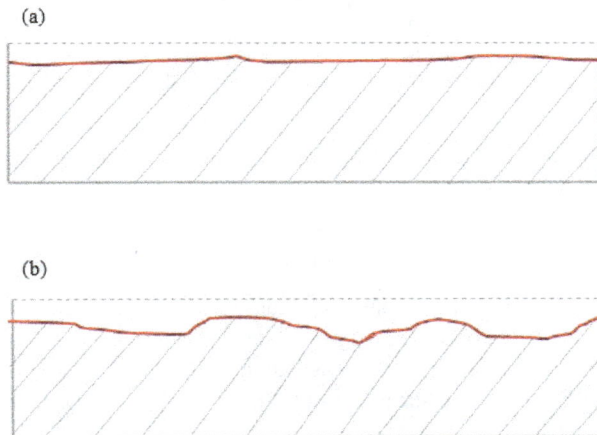

(a)

(b)

Figure 1.3.   Continuous corrosion: Uniform corrosion (a), Non-uniform corrosion (b).

The uniform corrosion occurs when the entire surface of the metal is exposed to a corrosive environment and is usually found in atmospheric corrosion of steel or corrosion of copper alloys in seawater [34].

1.2.5. Discontinuous corrosion

Discontinuous or localized corrosion occurs only on certain portions of the surface in contact with the corrosive environment [35]. This type of corrosion is divided into two categories:

- Macroscopic corrosion: points, spots and cavernous, galvanic, by cavitation (metal plucking);
- Microscopic corrosion: intergranular, under tension (intergranular, intragranular or mixed), filiform (cracking fatigue), selective (involves the preferential dissolution of a certain constituent of a metal alloy.

Galvanic corrosion occurs when two or more metals with different electrochemical potentials are in contact with an electrolyte. Also, the galvanic coupling can occur in the case of the same alloy, between its phases (phases, intermetallic compounds etc.) [36,37].

Pitting corrosion is defined as the result of surface corrosion of a metallic material, limited to a small area point, which takes the form of a cavity and has a width/depth ratio = 2/3 [38].

Pitting corrosion is one of the most destructive and intense forms of corrosion. It is considered more dangerous than widespread corrosion because it is more difficult to detect [39].

It can occur in any metal but is most common in metals that form protective oxide layers, such as aluminium or magnesium alloys. First, on the surface, a layer of white or grey powder is observed, similar to the dust that stains the surface of the metal. When this deposit is removed, small punctures or corrosion points can be observed on the surface. They can penetrate deep inside the metal and greatly damage the outer surface [36].

Among the materials in which we find this type of corrosion, are stainless steel, aluminium alloys, some copper alloys or Monel alloy (Ni-Cu alloy).

Depending on the mode of propagation of point corrosion in the material, several types of pitting are known that can be classified into true pitting (Fig. 1.3) and lateral pitting (Fig. 1.4).

The aggressive environments that can easily cause corrosion at the point are generally chlorine or bromine-based solutions. Aggressive anions can stimulate the development of corrosion at the point only when their concentration in the solution is higher than a certain critical value, which depends on the nature of the metal or alloy, the heat treatment and the condition of the surface.

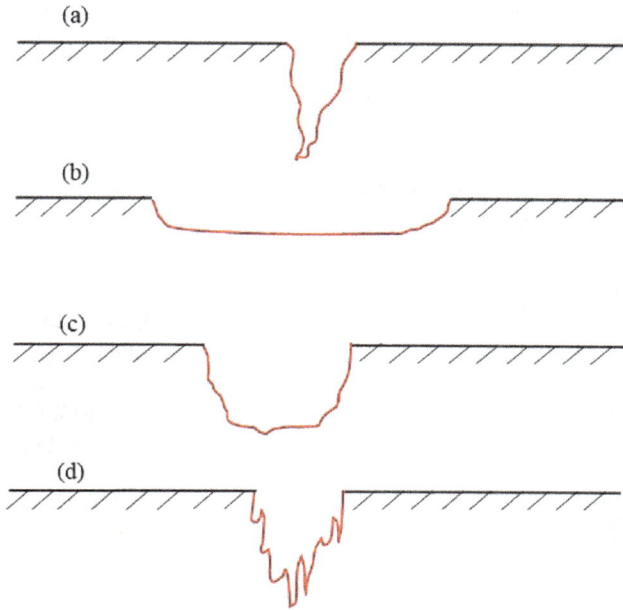

Figure 1.3.   Trough pits: Narrow, deep (a), Wide, shallow (b), Eliptical (c), Vertical (d).

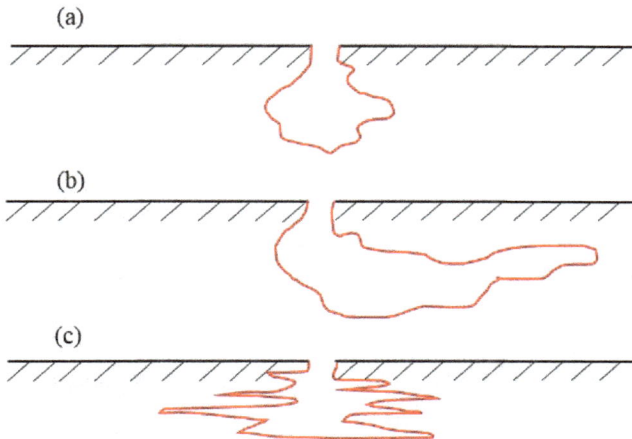

Figure 1.4.   Sideway pits: Subsurface (a), Undercutting (b), Horizontal (c).

Spot corrosion and intergranular corrosion are the most dangerous forms because without a change in the appearance of the metal, the part or subassembly is destroyed almost unexpectedly.

In Fig. 1.5. the influence of the corrosion type on the decrease of the resistance to breaking (R_m), in percent, of the duralumin is presented [25].

It is observed that for relatively small alloy losses, the change in the parameter (characteristic) is insignificant in the case of uniform generalized corrosion (a) and substantial for localized intergranular corrosion (c).

Localized corrosion occurs whenever there is a heterogeneity in the corrosive system considered, of the metal, of the environment or the physico-chemical conditions existing at the interface. At first glance the localized corrosion is generally visual, we traditionally distinguish the loosened corrosion, either in the form of the attack that results (point corrosion) or by localization (for example intergranular corrosion) [40]. In other cases, however, the appearance is associated with a mechanism (corrosion by crevice effect, erosion).

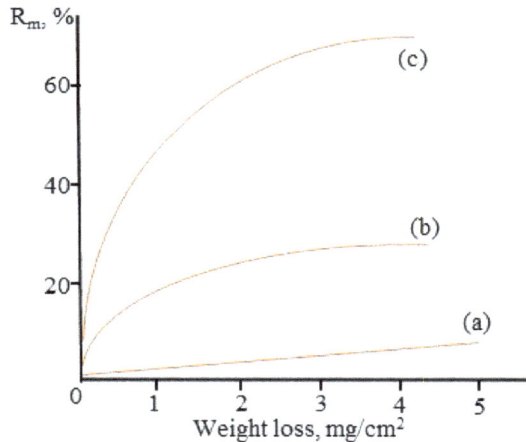

Figure 1.5. Modification of the resistance to breaking (R_m) of the duralumin depending on the type of corrosion: Uniform generalized corrosion (a), Pitting (b), Localized intergranular corrosion (c).

1.2.6. Intercrystalline corrosion

The intercrystalline corrosion (intragranular, Fig. 1.6) is characterized by the preferential destruction of the boundaries of crystalline grains weakening their cohesion, respectively the mechanical strength of the material. This is one of the most dangerous forms of destruction.

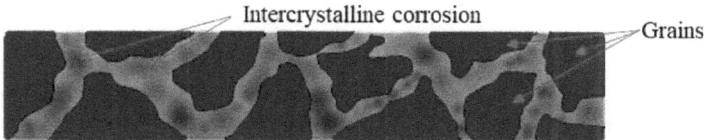

Figure 1.6.   Intercrystalline corrosion.

The attack takes place inside the metal, at the boundary between the grains and leads to a significant decrease in the mechanical and chemical properties of the material. In general, no changes are observed on the outer surface, which can lead to the unexpected destruction of the respective part.

Many corrosive environments can cause intercrystalline corrosion, including seawater, sulfuric acid, phosphoric acid, nitrate-based solutions or sulfonitric mixtures. Also, another factor that influences the appearance of intercrystalline corrosion is temperature.

Depending on the mode of propagation of intercrystalline corrosion, two types of corrosion can be distinguished:

- The intercrystalline corrosion under stress (corrosive cracking) - generally occurs in machine parts that work in corrosive environments and are subject to variable mechanical stresses. Pure metals are not susceptible to this type of corrosion
- Selective structure intercrystalline corrosion - occurs in low-alloy steels or sulfur cast iron with sulfur.

Intercrystalline corrosion can occur in aluminium alloys due to the segregation of impurities at the grain boundary or in stainless steels and some nickel alloys.

1.2.7. Transcrystalline corrosion

Transcrystalline (intergranular, Fig. 1.7) corrosion represents the destruction of the material by cracking the grains due to internal stresses that occur as a result of mechanical processing, heat treatment etc.

Transcrystalline corrosion   Grains

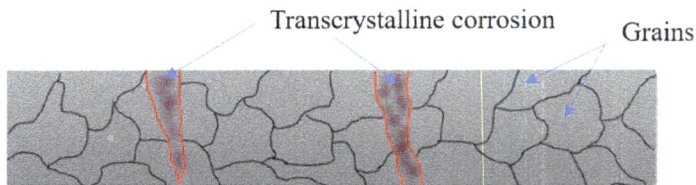

Figure 1.7.   Transcrystalline corrosion.

This is encountered in the form of trans-granular cracks.

1.2.8. Selective corrosion

Selective corrosion (Fig. 1.7) is achieved by attacking a certain constituent of the alloy. For example, the disengagement of brass in the acidic environment, following the selective dissolution of zinc, appears on the surface of the alloy a spongy layer of reddish-brown copper [33].

Selective corrosion

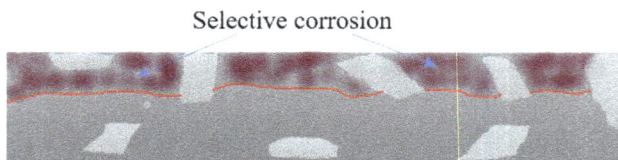

Figure 1.8.   Selective corrosion.

Brass with a content higher than 85% is more resistant to corrosion but is more expensive.

1.2.9. Atmospheric corrosion

Atmospheric corrosion is the most common and occurs under the effect of electrochemistry and the influence of thin films of moisture [41]. It depends on the area of existence: rural, urban, industrial, marine, etc. and meteorological factors: rain, dew, temperatures, wind etc.

The main atmospheric components that significantly influence corrosion are moisture (the main component on which causes corrosion of materials), oxygen (acts as a depolarizer) and carbon dioxide (the main component that conducts the film of moisture that condenses on the metal surface), but in addition to these, depending others can be added to the area of existence, which can lead to changes in the corrosion rate [14].

Depending on the existence, three important atmospheres are highlighted:

- Industrial atmosphere - containing sulfur compounds, which accelerates the corrosion process, increasing the corrosion rate;
- Marine atmosphere - which has high humidity and contains chlorides, chlorine ions penetrating the surface passivating layer;
- The rural atmosphere - does not contain foreign compounds, so the corrosion rate is low and the corrosion process is slow.

1.2.10. Underground corrosion

Corrosion in the soil is due to the action of the soil on the constructions embedded in it. Soils are different and considered porous colloidal systems. Corrosion increases substantially with increasing soil moisture. This is important because of the economic implications it has [15].

Example: corrosion of a crude oil transport pipeline incurs high costs for its replacement, respectively environmental pollution, production stoppage, etc. Also, for water pipelines, the costs are considerable for its replacement, restoration of the road or pavement.

The most important factors that lead to soil corrosion are the electrical resistance of the soil, its composition, as well as the permeability of the soil to water and air [42]. Depending on the permeability of the soil, the access of oxygen to the metal is limited or not leading to changes in the corrosion rate. For example, in the case of pipes passing under concrete ground, the corrosion rate will be higher than in a non-concrete area.

1.2.11. Aqueous corrosion

In distilled water, the corrosion process takes place due to the presence of oxygen and carbon dioxide. Therefore, if it is possible to limit the dissolved oxygen and carbon dioxide content, the aggressiveness of distilled water on the metal decreases by almost 0.

River water differs in terms of dissolved oxygen content. For example, in the waters of the plains, the amount of dissolved oxygen is lower compared to the waters of the mountain areas where the amount of oxygen is higher, thus increasing the aggressiveness of the environment [16].

In terms of corrosion, seawater (saltwater) is the most aggressive type of natural water. The main factors that increase its aggression are salinity, oxygen content, the existence of microorganisms etc.

Cave water or mine water has a different composition depending on the soil, so the corrosion process will take place at different corrosion rates [43].

Condensate water, which occurs in chemical plants, is very close in chemical composition to distilled water. However, if it contains free oxygen or carbon dioxide it becomes corrosive.

1.2.12. Corrosion due to mechanical stress

Corrosion due to the mechanical stresses occurs by combining chemical-corrosive and mechanical stresses in the operation of machine parts or subassemblies,

Stress corrosion occurs under the simultaneous action of the corrosive environment and the static or residual stresses in the material of the parts, being specific, in general, to stainless steels. The stresses present in the material come from a large number of sources and can manifest themselves in the form of tensile or compressive stresses. Tensions near the flow point favour corrosion [18]. For example, this type of corrosion occurs in titanium, aluminium or stainless steel when in a chlorine-containing environment.

At high temperatures and pressures in the presence of concentrated alkaline solutions, *caustic fragility* may occur.

Fatigue corrosion occurs when, in addition to the action of the corrosive environment, the material is also subjected to dynamic stresses. In the presence of the corrosive environment, at each dynamic stress of the material, its properties change, and after a certain number of operating cycles, there are disturbances in the arrangement of the metal atoms [44].

This phenomenon depends on several parameters, among which: the chemical composition of the material, its structure, the number of cyclic stresses, etc. Copper alloys or stainless steels are susceptible to corrosion.

Erosion corrosion occurs when the material is subjected to an abrasive environment at the same time as a corrosive liquid or when impurities in the fluid or its high flow rate come into contact with the material. Due to the abrasion wear of the metal, its surface becomes unprotected due to the removal of the protective layers, thus being exposed to the corrosive agent [45].

The conditions that favour the appearance of this type of corrosion are the high speed of liquid flow, impurities, turbulent flow, the way of designing the metal (elbows of a pipe), etc. The metals that form an adherent and hard layer that resist this type of corrosion are titanium and aluminium.

1.2.13. Microbiological corrosion

Microbiological corrosion is the destruction of the metal due to microorganisms that can occur under aerobic or anaerobic conditions. It occurs in the atmosphere, in the soil, in water and attacks everything that exists in an environment [46].

Corrosion processes are primed and stimulated in 60% of cases by biological elements: bacteria, actinomycetes, fungi, algae etc.

This type of corrosion is important when aluminium or copper steels or alloys are in contact with neutral water, especially stagnant water. Microbiological corrosion occurs in the form of spot corrosion or if there are colonies of sulfur-reducing bacteria, corrosion in crevices can also occur.

This corrosion usually occurs at water storage tanks or the bottom of pipes when water accumulates.

1.3. Metal passivation

Passivation is a complex chemical or electrochemical phenomenon, which consists of the qualitative modification of the metal-solution interface (corrosive environment), due to the formation of a surface oxide or a layer of adherent, continuous chemisorbed oxygen, which isolates the material from the corrosive environment [47].

Thus, a kinetic barrier without ionic conductance is formed between the material/solution phases, which opposes the corrosion process. The result of passivation is called passivity, ie the state of high corrosion resistance of metallic materials.

Passive transition differs from one metal to another. The following metals are most easily passivated, regardless of the nature of the environment: chromium, molybdenum, aluminium, titanium, tantalum and niobium, and in contact with alkaline solutions iron, cobalt, magnesium and nickel are easily passivated [20].

Some metals are more difficult to passivate and pass into a passive state only in strongly oxidizing environments, such as sulfuric acid, nitric acid, perchloric acid, etc. One such metal is lead, which is passivated into sulfuric acid, forming a protective layer of lead sulfate. Also, the silver in contact with hydrochloric acid passivates, forming a film of silver chloride.

The particularities of the behaviour of the metallic material that is passivated are studied following the dependence between the electrode potential and the density of the anodic current. Fig. 1.9 shows the anodic polarization curve, under potentiostat conditions.

Starting from the stationary potential E_s the curve shows the anodic dissolution of the metal according to the reaction eq. 1.6:

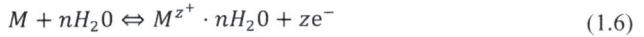

$$M + nH_2O \Leftrightarrow M^{z+} \cdot nH_2O + ze^-$$ (1.6)

where: M is the metal.

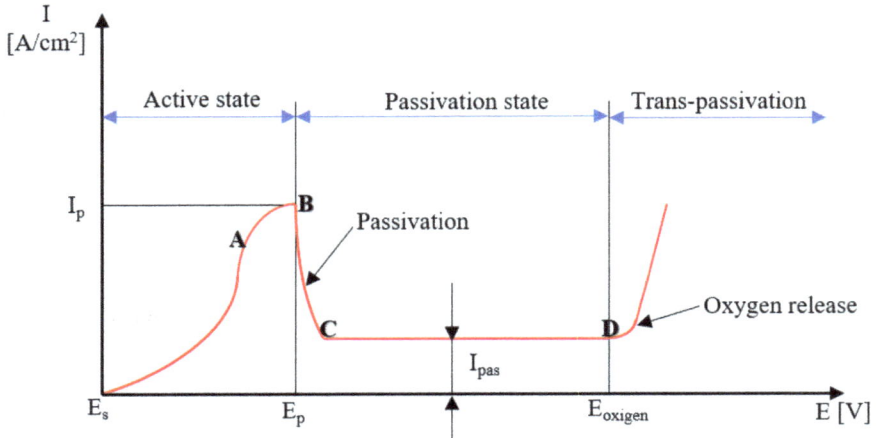

Figure 1.9.   Schematic representation of electrochemical passivation [25]

where: I_p – passivation current density;

E_p – passivation potential;

E_{oxygen} – the potential for oxygen release;

E_s – stationary potential.

At *A* the oxide protective layer begins to form, following the reaction eq. 1.7:

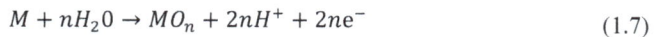

$$M + nH_2O \rightarrow MO_n + 2nH^+ + 2ne^-$$ (1.7)

Thus, the braking of the anodic dissolution process begins and the curve deviates from the simple logarithmic dependence.

In *B*, the tendency to increase the speed of the metal dissolution process is equal to the braking tendency of this process, due to the formation of the protective film. After *B*, the speed of formation of the protective film exceeds the dissolution rate of the metal and thus the anodic current decreases at a small displacement of the potential to higher values.

At C, the entire surface is covered with an oxide film, the pasty state is fully installed and the speed of the anodic process no longer depends on the potential, remaining at very low values.

After reaching the oxygen release potential, point D, the anodic polarization curve shows an increase due to the discharge of hydroxyl ions, the metal is subjected to the transpasivation process, respectively it becomes soluble again (corrodible).

The parameters that characterize the passivation state are the density of the critical passivation current, the passivation potential and the size of the passivation domain. For practice, it is useful to keep the value of the dissolution current density as low as possible and the passivation range as high as possible. The value of the passivation potential, also called Flade potential, varies from metal to metal and also depends on pH.

The passivation potential behaves like the oxide electrode potential, its value often corresponding to the normal oxide electrode potential.

In the case of passivation in oxidizing solutions, in addition to the anodic dissolution curve, the oxidant reduction curve must be taken into account, ie the equilibrium redox potential of the solution must be higher than the critical passivation potential and the speed of the cathodic reduction process. the oxidizing agent, at the passivation potential, is greater than the density of the critical passivation current [48].

Spontaneous passivation of a significant number of mats in contact with the atmosphere or various aggressive solutions leads to significant savings.

The ability to bring the metal into a passive state by anodic polarization, by introducing oxidants into the corrosive environment or by alloying offers convenient corrosion protection, along with many others.

References

[1]   Perez, N. Electrochemistry and Corrosion Science: Second Edition. *Electrochemistry and Corrosion Science: Second Edition* **2016**, 1–455. https://doi.org/10.1007/978-3-319-24847-9

[2]   Burduhos-Nergis, D.P.; Bejinariu, C.; Sandu, A.V. *Phosphate Coatings Suitable for Personal Protective Equipment*; Materials Research Foundations, 2021; Vol. 89; ISBN 9781644901113. https://doi.org/10.21741/9781644901113

[3]   Cai, Y.; Zhao, Y.; Ma, X.; Zhou, K.; Chen, Y. Influence of Environmental Factors

on Atmospheric Corrosion in Dynamic Environment. *Corrosion Science* **2018**, *137*, 163–175. https://doi.org/10.1016/j.corsci.2018.03.042

[4]   Edeleanu, C.; Littlewood, R. Thermodynamics of Corrosion in Fused Chlorides. *Electrochimica Acta* **1960**, *3*, 195–207. https://doi.org/10.1016/0013-4686(60)85003-7

[5]   El-Meligi, A.A. Corrosion Preventive Strategies as a Crucial Need for Decreasing Environmental Pollution and Saving Economics. *Recent Patents on Corrosion Science* **2010**, *2*, 22–33. https://doi.org/10.2174/1877610801002010022

[6]   Corrosion: Metal/Environment Reactions - Google Books Available online: https://books.google.ro/books?hl=en&lr=&id=FBH-BAAAQBAJ&oi=fnd&pg=PP1&dq=Corrosion+of+metallic+materials+is+due+to+their+thermodynamic+instability+in+relation+to+their+oxidized+form,−so+it+is+a+natural+phenomenon+by+which+metals+tend+to+return+to+a+more+stable+state,+lower+energy+of+oxides,+chlorides,+sulfates+&ots=yX2p_xow4b&sig=PR Jt3S8_sQGb5ciW3k34glWkQho&redir_esc=y#v=onepage&q&f=false (accessed on 17 September 2021).

[7]   TOMASHOV, N.D. Development of the Electrochemical Theory of Metallic Corrosion. *Corrosion* **1964**, *20*, 7t–14t. https://doi.org/10.5006/0010-9312-20.1.7t

[8]   Fundamentals of Electrochemical Corrosion - Ele Eugene Stansbury, Robert Angus Buchanan - Google Books Available online: https://books.google.ro/books?hl=en&lr=&id=baHwfLpWpP8C&oi=fnd&pg=PR1&dq=electrochemical+corrosion&ots=LudIK-Ck0M&sig=Je8ezVLMHbdF1l0KakKGMN2GiZU&redir_esc=y#v=onepage&q=electrochemical%20corrosion&f=false (accessed on 17 September 2021).

[9]   O'Connell, M.; McNally, C.; Richardson, M.G. Biochemical Attack on Concrete in Wastewater Applications: A State of the Art Review. *Cement and Concrete Composites* **2010**, *32*, 479–485. https://doi.org/10.1016/j.cemconcomp.2010.05.001

[10]   González, J.A.; Andrade, C.; Alonso, C.; Feliu, S. Comparison of Rates of General Corrosion and Maximum Pitting Penetration on Concrete Embedded Steel Reinforcement. *Cement and Concrete Research* **1995**, *25*, 257–264. https://doi.org/10.1016/0008-8846(95)00006-2

[11]   Zhang, Z.; Deng, Y.; Ye, L.; Zhu, W.; Wang, F.; Jiang, K.; Guo, X. Influence of Aging Treatments on the Strength and Localized Corrosion Resistance of Aged Al–Zn–Mg–Cu Alloy. *Journal of Alloys and Compounds* **2020**, *846*, 156223.

https://doi.org/10.1016/j.jallcom.2020.156223

[12]   di Maggio, R.; Rossi, S.; Fedrizzi, L.; Scardi, P. ZrO_2-CeO_2 Films as Protective
       Coatings against Dry and Wet Corrosion of Metallic Alloys. *Surface and Coatings
       Technology* **1997**, *89*, 292–298. https://doi.org/10.1016/S0257-8972(96)03014-9

[13]   Zheng, S.Q.; Chen, C.F.; Chen, L.Q. Corrosion Characteristics of 2205 Duplex
       Stainless Steel in High Temperature and High Pressure Environment Containing
       H_2S/CO_2. *Applied Mechanics and Materials* **2012**, *236–237*, 95–98.
       https://doi.org/10.4028/www.scientific.net/AMM.236-237.95

[14]   Leygraf, C.; Wallinder, I.O.; Tidblad, J.; Graedel, T. Atmospheric Corrosion:
       Second Edition. *Atmospheric Corrosion: Second Edition* **2016**, 1–374.
       https://doi.org/10.1002/9781118762134

[15]   underground corrosion: Part I: Corrosion Mechanism of Metals in Soil. *Anti-
       Corrosion Methods and Materials* **1958**, *5*, 5–9. https://doi.org/10.1108/eb019420

[16]   Robertson, J. The Mechanism of High Temperature Aqueous Corrosion of
       Stainless Steels. *Corrosion Science* **1991**, *32*, 443–465.
       https://doi.org/10.1016/0010-938X(91)90125-9

[17]   Loto, C.A. Microbiological Corrosion: Mechanism, Control and Impact—a
       Review. *The International Journal of Advanced Manufacturing Technology 2017
       92:9* **2017**, *92*, 4241–4252. https://doi.org/10.1007/s00170-017-0494-8

[18]   Woodtli, J.; Kieselbach, R. Damage Due to Hydrogen Embrittlement and Stress
       Corrosion Cracking. *Engineering Failure Analysis* **2000**, *7*, 427–450.
       https://doi.org/10.1016/S1350-6307(99)00033-3

[19]   Corrosion Science and Technology - 3rd Edition - David E.J. Talbot - Available
       online: https://www.routledge.com/Corrosion-Science-and-Technology/Talbot-
       Talbot/p/book/9780367735340 (accessed on 17 September 2021).

[20]   Luca, C.; Stratula-Vahnoveanu, B. *Chimie Generală*; Editura „Gheorghe Asachi":
       Iasi, 2003; ISBN 973-621-033-2.

[21]   Veprek-Siska, J. The Role of Metal Ions in Oxygen Activation . *Acta Biol Med
       Ger.* **1979**, *38*, 357–361.

[22]   Urdas, V. Tratamente Termice, Termochimice, Coroziunea Metalelor Si Acoperiri
       de Suprafata; Univ: Sibiu, 2001; ISBN 9736512991.

[23]   Ahmad, Z. *Principles of Corrosion Engineering and Corrosion Control*; Elsevier
       Ltd, 2006; ISBN 9780750659246. https://doi.org/10.1016/B978-075065924-

6/50004-0

[24] Smallman, R.E.; Ngan, A.H.W. Oxidation, Corrosion and Surface Engineering. *Modern Physical Metallurgy* **2014**, 617–657.https://doi.org/10.1016/B978-0-08-098204-5.00016-X

[25] Udrescu, L. *Tratamente de Suprafata Si Acoperiri*; Politehnica: Timisoara, 2000; ISBN 9739389740.

[26] Ahmad, Z. Types of corrosion: Materials and Environments. *Principles of Corrosion Engineering and Corrosion Control* **2006**, 120–270. https://doi.org/10.1016/B978-075065924-6/50005-2

[27] LEFRANCOIS, P.A.; HOYT, W.B. Chemical Thermodynamics of High Temperature Reactions In Metal Dusting Corrosion. *Corrosion* **1963**, *19*, 360t–368t. https://doi.org/10.5006/0010-9312-19.10.360

[28] Udrescu, O.V.; Plaiasu, G.; Barbulescu, C.A.; Ducu, M.C.; Moga, S.G.; Negrea, D.A. Corrosion of Some Military Equipment and Ustensils Discovered in the Ancient Site of Buridava.; 13th International Conference on Electronics, Computers and Artificial Intelligence (ECAI), August 23 2021; pp. 1–6. https://doi.org/10.1109/ECAI52376.2021.9515186

[29] Buchanan, R.A.; Stansbury, E.E. Electrochemical Corrosion. *Handbook of Environmental Degradation of Materials: Second Edition* **2012**, 87–125. https://doi.org/10.1016/B978-1-4377-3455-3.00004-3

[30] Sannier, J.; Flament, T.; Terlain, A. Corrosion of martensitic steels in flowing Pb17Li. *Fusion Technology 1990* **1991**, 901–905. https://doi.org/10.1016/B978-0-444-88508-1.50165-1

[31] Kadukova, J.; Pristas, P. Biocorrosion—Microbial Action. *Encyclopedia of Interfacial Chemistry: Surface Science and Electrochemistry* **2018**, 20–27. https://doi.org/10.1016/B978-0-12-409547-2.13500-0

[32] Usher, K.M.; Kaksonen, A.H.; Cole, I.; Marney, D. Critical Review: Microbially Influenced Corrosion of Buried Carbon Steel Pipes. *International Biodeterioration and Biodegradation* 2014, *93*, 84–106. https://doi.org/10.1016/j.ibiod.2014.05.007

[33] Florescu, A.; Bejinariu, C.; Comaneci, R.; Danila, R.; Calancia, O.; Moldoveanu, V. *Stiinta Si Tehnologia Materialelor*; Ed. Romanul: Bucuresti, 1997; Vol. II; ISBN 9739180469.

[34] Asri, R.I.M.; Harun, W.S.W.; Samykano, M.; Lah, N.A.C.; Ghani, S.A.C.; Tarlochan, F.; Raza, M.R. Corrosion and Surface Modification on Biocompatible

Metals: A Review. *Materials Science and Engineering: C* **2017**, *77*, 1261–1274. https://doi.org/10.1016/j.msec.2017.04.102

[35]   Luo, C.; Albu, S.P.; Zhou, X.; Sun, Z.; Zhang, X.; Tang, Z.; Thompson, G.E. Continuous and Discontinuous Localized Corrosion of a 2xxx Aluminium–Copper–Lithium Alloy in Sodium Chloride Solution. *Journal of Alloys and Compounds* **2016**, *658*, 61–70. https://doi.org/10.1016/j.jallcom.2015.10.185

[36]   Hack, H.P. Galvanic Corrosion. *Reference Module in Materials Science and Materials Engineering* **2016**. https://doi.org/10.1016/B978-0-12-803581-8.01594-0

[37]   Sastri, V.S. (Vedula S.); Ghali, Edward.; Elboujdaini, Mimoun. Corrosion Prevention and Protection : Practical Solutions. **2007**, 557. https://doi.org/10.1002/9780470024546

[38]   Forms of Corrosion - Aircraft Corrosion Control | Aircraft Systems Available online: https://www.aircraftsystemstech.com/2019/04/forms-of-corrosion-aircraft-corrosion.html (accessed on 21 September 2021).

[39]   Frankel, G.S. Pitting Corrosion of Metals: A Review of the Critical Factors. *Journal of The Electrochemical Society* **1998**, *145*, 2186. https://doi.org/10.1149/1.1838615

[40]   Totta, P.A. Abstract: In-process Intergranular Corrosion in Al Alloy Thin Films. *Journal of Vacuum Science and Technology* **1998**, *13*, 26. https://doi.org/10.1116/1.568867

[41]   Syed, S. Atmospheric corrosion of materials. *Emirates Journal for Engineering Research* **2006**, *11*, 1–24.

[42]   Azoor, R.M.; Deo, R.N.; Birbilis, N.; Kodikara, J. On the Optimum Soil Moisture for Underground Corrosion in Different Soil Types. *Corrosion Science* **2019**, *159*. https://doi.org/10.1016/j.corsci.2019.108116

[43]   Jeon, B.; Sankaranarayanan, S.K.R.S.; Duin, A.C.T. van; Ramanathan, S. Atomistic Insights into Aqueous Corrosion of Copper. *The Journal of Chemical Physics* **2011**, *134*, 234706. https://doi.org/10.1063/1.3599090

[44]   Fleck, C.; Eifler, D. Corrosion, Fatigue and Corrosion Fatigue Behaviour of Metal Implant Materials, Especially Titanium Alloys. *International Journal of Fatigue* **2010**, *32*, 929–935. https://doi.org/10.1016/j.ijfatigue.2009.09.009

[45]   Levy, A. v. The Erosion-Corrosion Behavior of Protective Coatings. *Surface and Coatings Technology* **1988**, *36*, 387–406. https://doi.org/10.1016/0257-8972(88)90168-5

[46]   Vargel, C. Microbiologically Influenced Corrosion. *Corrosion of Aluminium* **2020**, 289–294. https://doi.org/10.1016/B978-0-08-099925-8.00024-7

[47]   Godding, J.S.W.; Ramadan, A.J.; Lin, Y.H.; Schutt, K.; Snaith, H.J.; Wenger, B. Oxidative Passivation of Metal Halide Perovskites. *Joule* **2019**, *3*, 2716–2731. https://doi.org/10.1016/j.joule.2019.08.006

[48]   Griffin, G.L. A Simple Phase Transition Model for Metal Passivation Kinetics. *Journal of The Electrochemical Society* **1984**, *131*. https://doi.org/10.1149/1.2115505

Chapter 2

Methods for Testing and Assessing Corrosion Behaviour

2.1. Quantitative (direct) analysis of the corrosion behaviour of materials

To characterize the behaviour of a metal (alloy) in relation to a certain aggressive environment, the notion of chemical stability or corrosion resistance is generally used.

The corrosion resistance of metallic materials in various corrosive environments is assessed by experimental determination of the corrosion rate, by measuring the thickness of the oxide layer (corrosion products) or by determining the variation of mechanical, electrical, optical properties etc. on specially executed samples from the tested material or by establishing the consumption of corrosive agent [1].

The corrosion rate represents the variation of the sample mass, as a result of the corrosion, per unit area and time. This may express weight loss when corrosion products can be removed from the surface of the sample (a common situation in practice) or increase the weight of the samples, by forming corrosion products whose removal would lead to errors by damaging the unattached metal (case oxidation of metals), but an increase that cannot be accurately assessed, because it is not possible to determine precisely the amount of metallic material transformed by corrosion, not knowing the chemical composition of the products formed [2].

Complete removal from the corroded surface of corrosion products without affecting the base material can be done either mechanically or with specific reagents (for example, from corroded carbon steel, the products can be removed with 10% H_2SO_4 solution, for 30 minutes at 25 °C or with a reagent composed of 20% NaOH solution and 200 g/l zinc powder, for 5 minutes at 85 °C).

Corrosion can also be expressed by the volumetric index of hydrogen or oxygen (gases accompanying corrosion) [3,4], so the corrosion rate can be determined with the relation eq 2.1:

$$v_{cor} = \frac{K_v \cdot A}{22400 \cdot Z} \qquad (2.1)$$

where: A - mass number of the metal;

Z - valence of the metal ion.

The volumetric index can be calculated with the relation eq. 2.2:

$$K_v = \frac{0.36V \cdot (b - h)}{S \cdot \tau \cdot T}$$
(2.2)

where: V- the volume of gas released during the determination, in cm^3;

p – barometric gas pressure, in mm col. Hg;

h – water vapour pressure at operating temperatures, in mm col. Hg;

S- sample surface, in cm^2;

τ – experimentation time, in hours;

T- absolute temperature, in K.

Corrosion resistance can also be expressed using the penetration index, according to the following relation (eq. 2.3):

$$P = \frac{0.76v_{cor}}{\rho} = \frac{\Delta\delta}{\tau}$$
(2.3)

where: ρ – material density;

$\Delta\delta$ – average reduction in the thickness of the metal subject to corrosion;

τ – chemical attack time.

Corrosion, in general, and intercrystalline corrosion, in particular, can be estimated by measuring the variation of the physical properties of the metal under the action of the aggressive environment [3–5]. Thus, the size of the corrosion index is calculated with the relation eq. 2.4:

$$K_f = \frac{F - F_1}{F} \cdot 100$$
(2.4)

where F and F_1 represent the values of the physical-mechanical quantities (electrical resistance, the reflective capacity of the surface, the resistance limit, the residual elongation, before, respectively after the corrosion of the sample.

In the case of uniform corrosion, this is the most widely used method of measuring the corrosion rate. The weight loss of the material is used to calculate the loss in thickness of the metal, assuming that it acted uniformly. In some cases, the measurement of the thickness of the samples is also used, and the results are expressed in μm / year. The corrosion rate is usually calculated by weight loss and not by loss of sample thickness.

Intercrystalline corrosion can also be determined by the weight loss method. For example, the ASTM 262-86 standard indicates the procedure by which this method is performed (Streicher test), the samples being introduced for 24 hours at 120 °C in a solution based on iron sulphate and 50% sulfuric acid. This method measures the sensitivity of stainless steel and nickel-based alloys to intercrystalline corrosion due to the precipitation of chromium particles at the grain boundary [6,7].

The Huey Test is also used to determine the intercrystalline corrosion of chromium-nickel steels. It consists of introducing the samples five times, for 48 hours, in a solution based on nitric acid of 65% concentration. The corrosion rate during each period is determined by the weight loss of the samples. This test is used to detect chromium depleted areas, intermetallic precipitates (sigma phase in the material), as well as to check for stabilization or reduction of carbon content [8,9].

2.2. Indirect analysis of the corrosion behaviour of materials

The corrosion behaviour of materials can be evaluated indirectly by several methods:

- optical, when the examination is made with the naked eye or with a magnifying glass, establishing the type of destruction (generalized or localized corrosion) and microscopic, highlighting changes in the structure of the material by intercrystalline or transcrystalline corrosion [10,11];
- with SEM microscopy, when determining the difference in composition in the passive or corrosion film [12];
- with radioactive isotopes, identifying the type of ions that form in the corrosion film and the presence of anions in the solution [13];
- spectral analysis, which determines the structure and composition of the passivation or corrosion film [14];
- acoustic and electrical (variation of electrical resistance) [15];
- reflection capacity: in stainless steel the formation of the oxide film reduces the reflection, increasing the corrosion resistance [16];
- electrochemical, which measures the intensity of the corrosion current, the stationary potential of - corrosion, polarization resistance, electrochemical impedance, etc., draws polarization curves [17–19];
- mechanical, which involve measurements before and after corrosion at different stresses, such as bending, stamping, traction, fatigue etc.

2.2.1. Corrosion assessment by non-destructive methods

Visual testing also known as visual inspection is one of the most common techniques by which surface defects of a material can be observed. This can be helped by using optical instruments, such as magnifiers or computer-aided systems (known as "remote viewing"). This method allows the detection of corrosion, misalignment, damage, cracks and more. Visual testing is inherent in most other types of non-destructive testing because it depends very much on the experience of the person studying the surface of the material [20,21].

In the case of point/spot corrosion, the depth of the holes is measured. In some cases, it is considered the deepest, and in others, an average of the 10 deepest holes is made. In pressure vessels or containers, the effect of this type of corrosion can be much more severe than indicated by material loss. In the case of corrosion in crevices, the measurement of weight losses cannot be used, so the inlet diameter and the depth of the crevices are determined to determine the corrosion [22].

For intercrystalline corrosion, for example in the case of stainless steel, the Strauss test can be used, which consists in introducing the samples into a solution based on copper sulphate, sulfuric acid and distilled water at a temperature of 180 °C, at the end of which it is visually examined, according to ASTM A262 practice E [23,24].

To deepen the effects of corrosion processes, the surfaces of the materials can be observed under an optical microscope. For example, in Fig. 2.1 Corrosion points can be observed on the surface of carbon steel (C45 steel) [20].

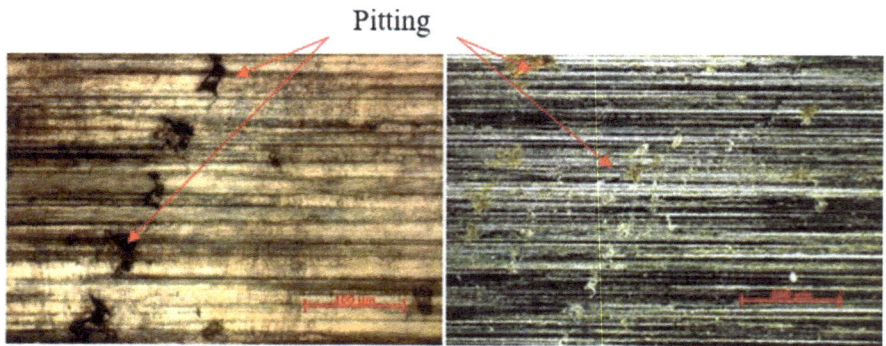

Figure 2.1. *Optical microstructure of carbon steel at magnification power: a) 40x; b)20x*

In addition, scanning electron microscopy (SEM) with integrated dispersed energy spectrometric analysis (EDS) can be used to provide us with additional information about the chemical composition structures of corroded surfaces and corrosion products. One such example is shown in Fig. 2.2, where the corroded surface of carbon steel immersed for 85 days in seawater was analyzed.

Figure 2.2.   SEM microphotographs for C45 immersed 24 hours in seawater a) 100x, b) 1000x and c) 5000x [17]

In Fig. 2.2 it can be seen that a thick layer of reaction products was deposited on the surface of the carbon steel immersed for 85 days in salt water, forming a crust that easily detaches from the surface. At a magnification of 5000X, some agglomerations of monoclinic crystals can be observed, which are characteristic of iron oxy-hydroxides. To confirm the type of crystals was studied using EDS analysis (Fig. 2.3) noting that the chemical composition of the analyzed surface consists of iron and oxygen, with small amounts of carbon, and can determine thus determining that the layer formed on the sample surface is FeO(OH) [17].

Laser profilometry uses a high-speed rotating laser light source and miniature optics to detect corrosion, pitting, erosion and cracks, detecting surface changes through a 3D image generated by the surface topography [21].

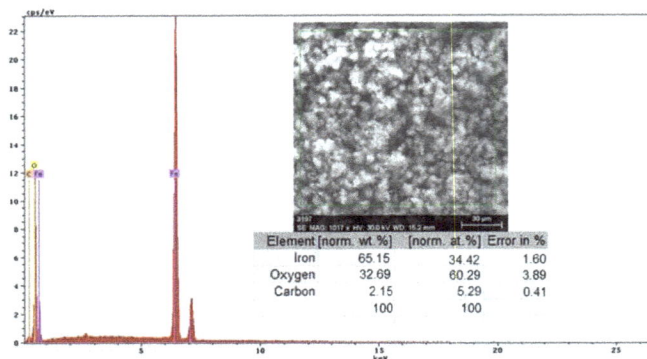

Figure 2.3.   Energy spectrum and crust composition on the C45 surface immersed for 85 days in seawater [17]

Alvarez et al. [25] studied the corrosion behaviour of the AE44 magnesium alloy in a NaCl-based solution, analyzing the samples by optical microscopy, SEM and laser profilometry. Because the weight loss method did not provide enough information, they used the laser profilometry method, analyzing the area and the volume of the pit according to the immersion time of the samples in solution, observing how the pit corrosion manifests. They also observed that as the immersion time increases, the pit becomes very shallow, while the intergranular corrosion allowed the eutectic area to degrade, thus the pits joining (Fig. 2.4).

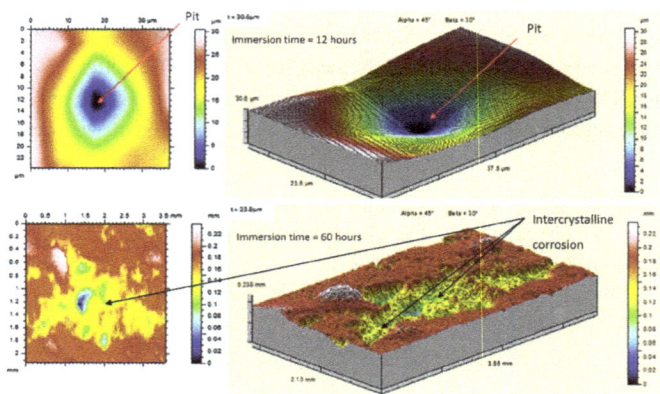

Figure 2.4.   Laser profilometry for AE44 magnesium alloy immersed in NaCl (from R.B. Alvarez et al. [25]*)*

Time of flight diffraction (TOFD) is the process of changing the wavelength of sound because it interacts with a discontinuity in a material. This mechanism is used in situations where true reflection cannot be obtained, but sufficient diffraction occurs to change the flight time of the sound in a gripping arrangement [21,26]. This method is used to detect the tip of a defect that is perpendicular to the contact surface of the samples. ToFD is also used for rear wall inspection to detect corrosion [26,27].

Gordon et al. [28] studied the painted and corroded surfaces of pieces taken from an out-of-service bridge, analyzing the influence of surface conditions (corroded or painted surfaces) testing the accuracy of the TOFD method to detect and measure surface defects. They concluded that corroded surfaces reduce the amplitude of ultrasonic signals, but do not affect the ability of the method to detect and size defects. It was also observed that depending on the amplitude signal, the degree of surface corrosion can be approximated (signal amplitude of the order of 10% - slightly corroded surfaces while for strongly corroded surfaces the signal amplitude reductions will be greater than 25%).

Ellipsometry is a non-destructive method that can determine the corrosion behaviour of materials and is based on the property of thin layers to change the polarization state of light reflected at angles less than 90° from a polarized light in the plane to a polarized one. Elliptical [29,30].

Poksinski et al. [31]demonstrated that corrosion monitoring can be performed through internal reflection ellipsometry, a technique that combines ellipsometry with total internal reflection mode. They performed dynamic and static tests on thin layers of copper. In the case of the static test, the copper layers were exposed for long periods to hydrochloric acid (7 hours). The ellipsometric spectrum was recorded for the samples immersed in HCl for one hour and 7 hours (Fig. 2.5).

It was observed that regardless of the type of test (dynamic or static) the effects of hydrochloric acid on the copper layers are similar in quality, and the final thickness of the layer after about one hour of exposure to HCl is 32.4 nm, the copper oxide layer being 5.7 nm thickness, while after 7 hours of exposure the layer thickness is 33.5 nm, the copper oxide layer having a thickness of 4.6 nm. Thus, Poksinski et al. [31]concludes that this method can be used to monitor corrosion, depending on the type of application and the type of metal.

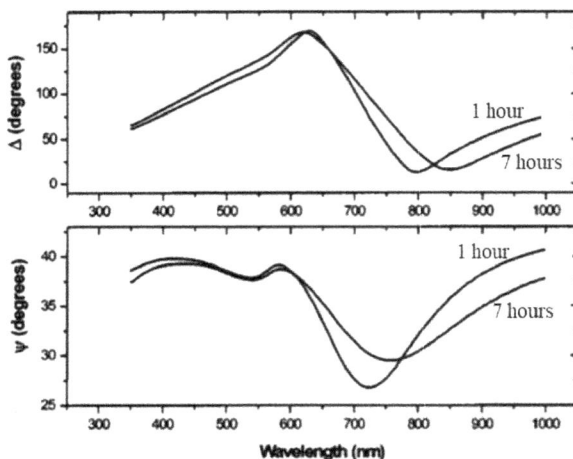

*Figure 2.5.   Ellipsometric spectra recorded for copper layers exposed to HCl for
different times (1 hour and 7 hours) (from Poksinski et al.* [31]*)*

2.2.2.   Corrosion assessment by electrochemical methods

Corrosion evaluation by linear polarization

The most natural corrosion processes take place through electrochemical reactions that take place in wet environments. For this reason, electrochemical study methods are a fast and efficient means of obtaining information on the thermodynamic probability of corrosion of metal immersed in a liquid, instantaneous corrosion rate (corrosion rate at the simple immersion of the metal in the corrosive environment), the type of corrosion, as well as the factors that may influence the corrosion (temperature, pH, accelerators or corrosion inhibitors) [20].

Electrochemical corrosion is an oxidation-reduction process in which anodic oxidation of the metal takes place simultaneously with a cathodic reduction process, in which the electrons generated in the anodic reaction are consumed [32]. The two processes are coupled and run at the same speed, at a common potential called mixed potential, or corrosion potential (E_{cor}). The speed with which the two processes take place is directly proportional to the intensity of the electric current.

The thermodynamic tendency of oxidation of a metal or alloy immersed in an electrolytic medium can be expressed by the corrosion potential, E_{cor}, while the corrosion rate is expressed by the intensity of electron transfer anode → cathode.

At the corrosion potential, the intensity of the anodic current (I_a) is equal to the intensity of the cathodic current (I_c), and the current in the external circuit is zero. Under these conditions, the metal/solution system is at thermodynamic equilibrium, $(I_a-I_c)_{E=Ecor}=0$ and on the surface of the apparent metal, there is no net reaction. However, metal degradation occurs only in the anodic process, by oxidation, so that corrosion occurs at a rate proportional to the intensity of the anodic current (equal to the cathodic current): $I_{cor}=(I_a)_{Ecor}=(I_c)_{Ecor}$ proportional to the corrosion current, I_{cor}, respectively current density, $j_{cor} = I_{cor}/S$ [33].

The current density is a measure of the speed with which the oxidation process takes place at equilibrium, by simply immersing the metal or alloy in solution (S represents the total area of the metal in contact with the solution).

The corrosion potential can be measured directly in relation to a reference electrode, immersed directly or using a salt bridge in the same solution as the studied metal. After reaching the potential corrosion balance, it is measured with a millivoltmeter with a very high input impedance [18].

A faster method for evaluating the corrosion potential is based on obtaining linear polarization curves and using Evans diagrams. To obtain a linear polarization curve, the electrode, the studied metal, is introduced into the corrosive environment together with an auxiliary platinum electrode using which the metal can be polarized. Voltages are applied between the two electrodes, as well as a reference electrode that is used to measure the potential of the metal at a given moment. The potential of the metal is varied in the vicinity of the corrosion potential, from values lower than $E_{cor} = E_0$ to values higher than E_{cor} (anodic potentiodynamic polarization) and the intensity of the current passing through the circuit (between the metal and the reference electrode) is measured [34].

In the Evans, diagrams are represented graphically the dependence of the logarithm of the current density as a function of the suprapotential applied to the electrode, both for the oxidation reaction of the metal and for the reduction of the depolarizer on the electrode. For relatively small suprapotentials ($\Delta E=E - E_{cor}$, de ordinal a $\pm50\div60$ mV) the two curves are practically linear in the E-logJ coordinate system, and the intersection of these lines gives on the axis of the potentials the value of the corrosion potential (Fig. 2.6).

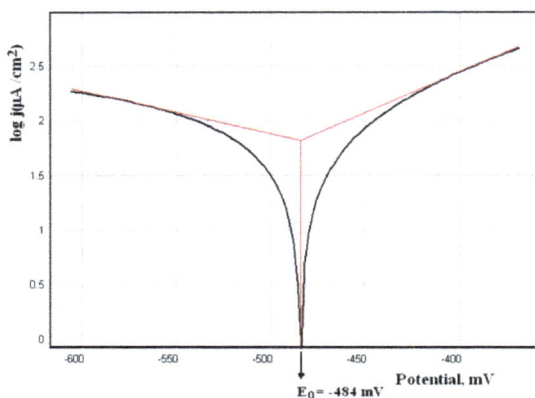

Figure 2.6. Evans diagram

a)   *The corrosion potential* is the electric potential measured in relation to a reference electrode, when the current in the external circuit is equal to zero, the reason for which it is also called potential in an open circuit [35]. This potential varies over time until a pseudo-stationary state is reached, which can be considered a steady-state. In some cases, the balance is reached only after 24 hours while in others after approx. two hours and even after 15-30 minutes.

The method based on the use of the Tafel diagram is dynamic because the electrode potential is always varied being more or less far from the equilibrium state (E_{cor}). The closer the equilibrium state is to the system, the lower the scanning speed (scanning) of the electrode potential (theoretically equal to zero). For this reason, the corrosion potential evaluated from the Tafel diagram will depend on the sweep speed. However, very low scanning speeds greatly prolong the determination time, during which time undesirable effects such as self-passivation can occur.

Variations in the open circuit potential over time provide qualitative information on the processes that take place at the material/solution interface. Open circuit potentials can become more positive or negative. Thus, the material becomes more susceptible to corrosion or passivates over time.

It should be noted, however, that the corrosion potential only expresses the corrosion tendency, ie it is a measure of the thermodynamic probability that the process takes place and does not take into account any factors that could slow down the process, generically called kinetic factors. For this reason, the assessment of the aggressiveness of a particular environment on an alloy only based on corrosion potential is inconclusive [36].

b)   *The instantaneous corrosion current*, I_{cor}, represents the current in the circuit, measured at the corrosion potential of the metal or alloy (E_{cor}), being the corrosion current that appears at the metal / corrosive environment interface when the metal is immersed in a solution. From a practical point of view, the density of the instantaneous corrosion current, $J_{cor}=I_{cor}/S$, is important, because it can be correlated directly with the corrosion rate [37,38].

The density of the instantaneous corrosion current can also be evaluated based on the linear polarization curve using the method of polarization resistance, eq. 2.5, representing the slope of the tangent to the potential-current density curve $[E = f(j)]$ at the equilibrium point ($E = E_0 = E_{cor}$ or $\eta = 0$, where $\eta = E - E_0$ is the suprapotential), ie at the potential corrosion [33].

$$R_p = \left[\frac{\Delta E}{\Delta j}\right]_{E=E_0} \tag{2.5}$$

The method is based on the evaluation of the polarization resistance, R_p, based on the Butler-Volmer equation (eq. 2.6) [39,40]:

$$j = j_{cor}\left[exp\left(\frac{2.303(E - E_{cor})}{b_a}\right) - \exp\left(-\frac{2.303(E - E_{cor})}{b_c}\right)\right] \tag{2.6}$$

where: b_a (eq 2.7) and b_c (eq. 2.8) represents the slopes of the linear portions of the anodic and cathodic branches in the $E = f(\log J)$ diagram, known as the Tafel slopes:

$$b_a = \left(\frac{\partial E}{\partial \log j}\right)_a = \frac{RT}{\alpha nF} \tag{2.7}$$

$$b_c = \left(\frac{\partial E}{\partial \log j}\right)_c = \frac{RT}{(1 - \alpha)nF} \tag{2.8}$$

It has been observed experimentally that j varies approximately linearly with the applied potential (E), starting from about $50 \div 60$ mV concerning the corrosion potential and only over a range of about $10 \div 20$ mV [7]. Stern and Geary [8,9] simplified the Butler-Volmer equation for the case of small overpotentials compared to E_{cor}, mathematically linearizing this equation by developing in series the logarithmic terms ($ex = 1 + x + x2/2! + x3/3! + ...$) and neglecting higher-order terms.

The simplified equation obtained can be transcribed in a simple form (eq. 2.9):

$$R_p = \left(\frac{dE}{dj}\right)_{(E_{cor})} = \frac{b_a \cdot b_c}{2.303 \cdot j_{cor} \cdot (b_a + b_c)} \quad (2.9)$$

After re-arranging the equation it can be obtained for instantaneous corrosion current (eq. 2.10):

$$j_{cor} = \frac{b_a \cdot b_c}{2.303 \cdot R_p \cdot (b_a + b_c)} \quad (2.10)$$

The polarization resistance can be expressed in absolute value or relative to the unit area depending on the size considered on the current axis (current intensity or density), the units of measure for resistance depending on the units of measure for E and I or j. If E is expressed in mV and I in mA, or E in V and I in A, then R_p is expressed in Ω. If E is expressed in V and j is expressed in A/cm^2, then R_p is expressed in $\Omega \cdot cm^2$.

Corrosion current density, calculated with Eq. 2.8 is obtained in mA/cm^2 if R' is expressed in $\Omega \cdot cm^2$ and b_a and b_c in mV/decade, or A/m^2, if R' is expressed in $\Omega \cdot cm^2$ and the Tafel constants in V/decade. The equations are also discussed in detail in the bibliographic references [41] and [42].

c)   *Instantaneous corrosion rate*

Gravimetric measurements determine the corrosion rates, which are generally very low, and can be expressed by the relations eq. 2.11 and eq. 2.12:

Superficial corrosion rate ($g/cm^2 \cdot an$):

$$v_s = 365 \frac{\Delta m}{S \cdot t} \quad (2.11)$$

Penetration rate ($\mu m/an$):

$$v_p = 10 \cdot 10^4 \cdot \frac{v_s}{\rho} \quad (2.12)$$

To correlate the corrosion rates defined by the above equations (v_s, respectively v_p) with the density of the instantaneous corrosion current (j_{cor}) Faraday's Law is used (eq. 2.113), which correlates the mass of separate substance at the electrode (deposited or removed) with the amount of electricity passed through the solution and the electrochemical equivalent of the substance:

$$\Delta m = k \cdot Q = k \cdot I_{cor} \cdot t = \frac{A \cdot I_{cor} \cdot t}{z \cdot F} \quad (2.13)$$

where: Δm – decrease the sample weight during time t;

I_{cor} – corrosion current;

$k = A/zF$ – the electrochemical equivalent of the corroding metal;

A - atomic mass;

z - the number of electrons involved in the oxidation process;

F = 96487 C/val – the Faraday number.

Dividing eq 2.11 by the product (area - time), we obtain the relation eq. 2.14:

$$\frac{\Delta m}{S \cdot t} = \frac{A \cdot I_{cor}}{z \cdot F \cdot S} \tag{2.14}$$

and taking into account the definition of the reaction rate and the definition of the current density $(j = I/S)$, it is obtained for the surface corrosion rate, v_s (g/cm²·s) (eq. 2.15):

$$v_s = \frac{A \cdot j_{cor}}{z \cdot F \cdot \rho} \tag{2.15}$$

The penetration speed, defined by the relation (eq. 2.12), is transformed according to the current density, replacing the expression of the surface corrosion rate, from the equation (eq. 2.15), when the relation eq. 2.16:

$$v_p = \frac{A \cdot j_{cor}}{z \cdot F \cdot \rho} \tag{2.16}$$

where: A/z - the electrochemical equivalent of the alloy metal or corrodible metal expressed in (g/val);

F = 96487 C/val;

ρ - the density of the metal or alloy (g/cm³),

j - current density (A/cm²).

Because the two speeds, v_s and v_p are generally very low, in practice more convenient units of measurement are used, taking into account longer time intervals.

Thus, the surface corrosion rate will be expressed in g/(cm²·year) or kg/(m²·year) with the relations eq. 2.17 and eq. 2.18

$$v_s = 0.327 \cdot \left(\frac{A}{z}\right) \cdot j_{cor} \ (g/(cm^2 \cdot an)) \tag{2.17}$$

$$v_s = 3.27 \cdot \overline{\left(\frac{A}{z}\right)} \cdot j_{cor} \ (kg/(cm^2 \cdot an)) \tag{2.18}$$

and the penetration speed (eq. 2.19):

$$v_p = 3.27 \cdot \overline{\left(\frac{A}{z}\right)} \cdot \frac{j_{cor}}{\rho} \ (mm/an) \tag{2.19}$$

In the case of alloys, the chemical equivalent will have an average value, calculated with the relation eq. 2.20:

$$\overline{\left(\frac{A}{z}\right)} \cdot \Sigma g_i \frac{A_i}{z_i} \tag{2.20}$$

where: g_i represents the gravimetric fractions (mass fractions) of the alloy components (A_i/z_i) the corresponding chemical equivalents.

Corrosion evaluation by cyclic polarization

One of the methods for further characterization of corrosion processes is cyclic potentiodynamic polarization. To obtain the cyclic polarization curves, also called cyclic voltammograms, the polarization of the studied alloy is performed continuously, with a known potential scanning speed (mV/s) and the current in the circuit is recorded. Sweeping the potential and recording the current variation is done automatically, obtaining a continuous curve. The speed of variation of the working electrode potential is indicated to be high enough to obtain current intensities high enough to cover possible accidental fluctuations in the system, but low enough to detect all the processes that could have a place in solution or on the surface of the electrode [43,44].

From the analysis of cyclic polarization curves can be obtained information on the type of electrochemical process that takes place on the surface of the alloy placed in the electrolytic environment (such as generalized corrosion, localized corrosion, passivation reduction-oxidation of species in solution) [38], evaluation of characteristic potentials corrosion, breakthrough potential, re-passivation potential, protection potential). Based on the currents recorded at potentials other than the equilibrium potential E_0, the corrosion rate under load can be calculated (when a higher potential than the corrosion potential is applied to the metal or alloy immersed in the solution) [45,46].

Depending on the purpose, the experimental data are represented in several ways, more often used is the representation $j = f(E)$ and the semi-logarithmic representation: $E = f(\log j)$, E being the potential applied to the alloy, and j - being the density total current at that potential. Fig. 2.7 illustrates two types of representation of a cyclic voltammogram using the experimental data obtained for a carbon steel sample coated with a zinc phosphate layer immersed in salt water [20].

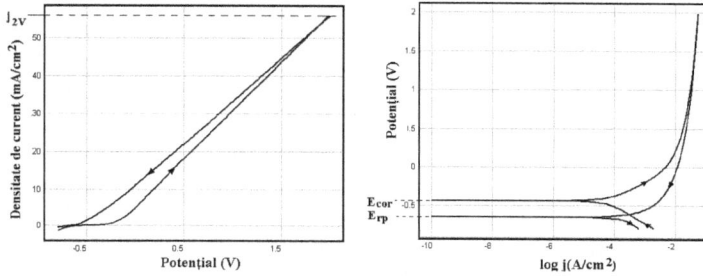

Figure 2.7.   Cyclic polarization curve and the assessment of the parameter's corrosion process

The characteristic quantities that can be evaluated from this curve are corrosion potential (E_{cor}) and repassivation potential (E_{rp}).

The corrosion potential (E_{cor}) is the potential at which the metal or alloy passes from a passive state, in which no oxidation process takes place, to an active state when the corrosion of the metal begins. The value of the corrosion potential evaluated in this diagram is not equal to the corrosion potential evaluated in the linear polarization curve. The difference is due to the fact that the corrosion potential evaluated from the linear polarization curve is very close to the equilibrium value as the scanning speed of the potential is very low (0.5 mV/s) while in the case of the cyclic voltammogram the scanning speed is high. (10 mV/s) and the system is far from equilibrium. The corrosion potential corresponds to the potential at which the anodic branch of the cyclic voltammogram changes from negative values to positive values of the current density [17].

The repassivation potential (E_{rp}) is the potential below which all active corrosion points are repassivated. Below this potential value the metal or alloy is passive (no longer corrodes). The repassivation potential corresponds to the potential at which the cathodic branch of the cyclic voltammogram passes from positive values to negative values of the current density.

Both the corrosion potential and the repassivation potential are determined more precisely from the semi-logarithmic diagram as can be seen in Fig. 2.7.

The shape of the cyclic voltammograms and the position of the anodic and cathodic branches allow obtaining information on the type of corrosion (generalized corrosion, point corrosion, passivation etc.) [18].

Based on these curves, the characteristic parameters of the processes that take place when applying a relatively high potential on the alloy can be evaluated. The study is important because this process accelerates the processes that can take place on the surface of the alloy immersed in the corrosive environment and thus can predict the behaviour of the alloy if it were immersed for a long time in solution [19,20].

In the case of voltammograms that in certain potential domains have a linear variation of the current density depending on the potential, the equations of the linear segments on the anodic and/or cathodic branch were also evaluated: $j = a \cdot E + b$. Because in some cases on the anodic branch (direct polarization curve; from negative values to positive values) to a certain value of the current density appears a jump, in the tables are presented only the linear equations of the cathodic branches.

Corrosion determination by electrochemical impedance spectroscopy

Electrochemical impedance spectroscopy is an electrochemical method, in alternating current, to study the processes that take place in a galvanizing cell. This method is used to characterize the processes at the electrode-electrolyte interface [47].

The disturbance of the electrochemical systems is achieved by applying an alternative signal of low amplitude (> 10mV), observing that the electrode returns to a steady state. This small alternative signal is an advantage of the method, as it offers the possibility to make repeated determinations on the same sample without changing its properties [48].

When we talk about metals (alloys), the main processes that take place at the electrode-electrolyte interface are the process of corrosion (or oxidation), so on the surface of the metal is formed a layer called an electric weak layer. It helps to slow down the corrosion process, changing the corrosion rate, thus being the kinetic factor of the process (reaction).

In the case of covering the metal surface with other protective layers, another process that occurs at the electrode-electrolyte interface is the partial degradation of the deposited layers (detachment from the metal surface, pore enlargement, change of layer thickness etc.) [49].

The corrosion rate depends very much on the layers deposited on the surface of the material or on the layers that appear as a result of the chemical reactions between the electrode and the electrolyte. Thus, the corrosion process depends on the kinetic factor and the diffusion factor (diffusion of ions involved in the corrosion process) [50,51].

One of the most important methods for studying corrosion processes is Electrochemical Impedance Spectroscopy because the information provided by this method makes it possible to study the mechanism of corrosion processes, quantitative determination of the protective capacity of a layer deposited on the metal surface and its degradation. With the help of the data obtained by EIS, an equivalent electrical circuit is established which is correlated with the metal/layer and layer/solution interfaces and by the phenomena that are manifested in the passive film [52,53].

By this method, it is possible to study the surfaces with high electrical impedance, which is indicated for the evaluation of the degradation of some organic layers deposited on the metal surface.

In electrochemical impedance measurements, a perturbation is applied to a corrodible metal in the form of a sinusoidal potential, E (t), and the system response, I (t) is recorded (eq. 2.21 and eq. 2.22) [54].

$$E = E_m \sin(\omega t) \tag{2.21}$$

$$I = I_m \sin(\omega t + \varphi) \tag{2.22}$$

where: E_m – maximum potential;

ω – angular frequency ($\omega = 2\pi f$, f –Hz);

I_m – maximum current;

φ – phase angle (phase shift) between current and potential.

The equipment used records the current-time and potential-time curves as a function of frequency, based on which the impedance spectrum is obtained.

In alternating current, the impedance is the equivalent of the resistance in direct current. For the relation between current (I) and voltage (E), in direct current Ohm's law is used: $E = I \cdot R$ (R being the resistance), while in alternating current the relation $E = I \cdot Z$ (Z being circuit impedance). Unlike resistance, the impedance of a circuit depends on the frequency of the applied signal: $Z(\omega) = E(\omega)/I(\omega)$, E and I being expressed by the relations eq 2.21 and eq. 2.22.

$Z(\omega)$ is a complex vector quantity (Fig. 2.8) having a real and an imaginary component, whose values also depend on the frequency (eq. 2.23):

$$Z(\omega) = Z'(\omega) + j^2 \cdot Z''(\omega) \tag{2.23}$$

where: $Z'(\omega) \equiv Zreal = |Z(\omega)| cos(\theta)$ is the real component of impedance

$Z''(\omega) \equiv Zim = |Z(\omega)| \sin(\theta)$ – is the imaginary component of impedance,

j^2 – the square of the imaginary number , or –1, $|Z(\omega)| = (Z'(\omega) 2 + Z''(\omega)2)1/2$

θ – is the phase shift (phase angle)

The inverse of the impedance is called the admittance and is usually denoted by Y ($Y = 1/Z = I/E$)

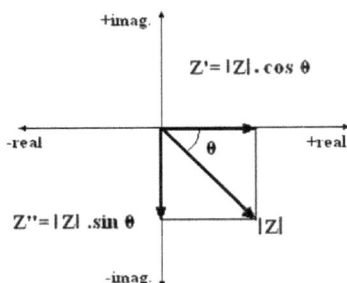

Figure 2.8.   Impedance vector in Cartesian coordinates

The definition of a correct equivalent electrical circuit for the analyzed system can be done only if the impedance dependence for the corrodible system is known.

The equivalent electrical circuit obtained contains the following elements: resistors, inductors, capacitors etc.

Within the electrical circuit obtained, each electrical element in the circuit corresponds to physical or chemical properties of the analyzed system, the most important being:

- R_s - the resistance of the solution between the reference electrode and the measuring electrode;
- C_{dl} – the capacity of the double-electric layer at the electrode/electrolyte interface;
- R_p - polarization resistance;
- R_{ct} – charge transfer resistance (ct- charge transfer);
- W – Warburg impedance is an impedance that occurs when a diffusion process opens at the solution electrode interface;
- C_{ext} – coating capacity;
- Q – constant phase element - which is introduced when in EIS experiments the capacitors do not behave ideally.

The constant phase element Q is introduced in the case of dielectric losses. When dielectric losses occur at the metal/electrolyte interface and the dielectric permittivity becomes a function of frequency and is represented by a complex number [47,55].

Warburg impedance is introduced when the diffusion of active species from the solution to the metal surface and vice versa is high enough and can influence the speed of the corrosion reaction.

The data obtained by the EIS are represented in the form of two types of diagrams [17,19]:

- Nyquist diagram - the abscissa represents the real part of the impedance, and in the ordinate, the imaginary part of the measured impedance taken with changed sign (-ImZ vs. ReZ);
- Bode diagram - in which the impedance modulus is represented according to the frequency logarithm and the phase angle according to the frequency logarithm $|Z|$vs. Frequency, and θ vs. Frequency).

Determination of galvanic corrosion

When two metals or alloys are in contact in the presence of an electrolyte, a galvanic torque is formed if the two metals have different electrode potentials. The electrolyte is a medium for the movement of metal ions (M^+) and water ions (H^+ and HO^-). This causes the metal at the anode to corrode faster than in the absence of the coupling, and the corrosion of the metal at the anode (the nobler metal) is braked or even stopped. The presence of the electrolyte can cause corrosion in the couple, even when the metals alone do not corrode in that electrolyte. Thus, the coupling of two different metals has the effect of increasing the corrosion rate of the less noble metal (anode) and decreasing the corrosion rate of the nobler metal [56,57].

The reactions that take place are analogous to those that take place in the case of an uncoupled metal (the least noble) but take place at a considerably higher speed. When a metal corrodes, two processes take place simultaneously. One is the dissolution of the metal at the anode. In the case of ferrous alloys, the metal that corrodes is iron:

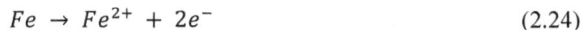

$$Fe \rightarrow Fe^{2+} + 2e^-$$ (2.24)

This reaction at the anode is compensated by a reaction at the cathode. In most cases of galvanic corrosion, the electrolyte also contains dissolved oxygen so that in alkaline media the primary cathodic reaction is the reduction of dissolved oxygen:

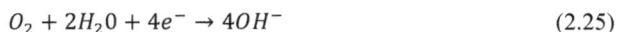

$$O_2 + 2H_2O + 4e^- \rightarrow 4OH^-$$ (2.25)

Materials Research Forum LLC
https://doi.org/10.21741/9781644901670

In acidic environments the cathodic reaction is often the reduction of hydrogen ions to hydrogen gas:

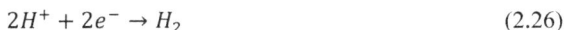

$$2H^+ + 2e^- \rightarrow H_2 \tag{2.26}$$

Other cathodic reactions may occur in aerated environments, such as partial oxidation of sulfur species.

In the case of corrosion of non-coupled metals, the anodic and cathodic reactions occur on small local areas on the metal surface, while in the case of bimetallic couples the cathodic reaction takes place almost entirely on the more electropositive metal of the couple, and the anodic reaction takes place on the more electronegative metal of the couple [58,59].

The parameters that characterize the galvanic corrosion of a pair of metals [57], the potential of the couple - E_{couple} and current density for the couple-j_{couple}) are influenced by several factors, such as:

- the difference between the electrode potentials of the couple partners;
- the quality of the electrical connection between metals;
- the electrolyte conductivity;
- the ratio of the exposed areas of the two electrodes;
- the distance between electrodes,
- the oxygenation degree of the electrolyte;
- the metallurgical composition etc.

Potential dynamic measurements and linear polarization curves can be used to predict and evaluate galvanic corrosion. Let be two different metals (alloys) whose polarization curves are shown in Fig. 2.9. The figure shows both hair curves and Evans lines for both materials. Metal 1 is the nobler metal (has the most positive value of the corrosion potential - E_{cor} (M1) and has a lower corrosion rate when not coupled. If the exposed surfaces of the two metals are equal and they are coupled then the metal interface/solution for both metals are at the same potential. The intersection between the cathodic branch of the nobler metal (M1) and the anodic branch of the nobler metal (M2) gives the couple potential – E_{couple} (measured relative to the reference electrode) and the density of current – j_{couple} when the two metals are immersed in the same electrolyte.

The potential of the nobler metal has been shifted from its E_{cor} value (Metal 1 – M1) to more negative values, which means a lower dissolution rate. The potential of the more active metal (Metal 2 - M2) is shifted from the E_{cor} value (M2) to more positive values leading to a higher dissolution rate.

The couple current density, j_{couple}, can be used to calculate the increase in the corrosion rate of the metal M2. Due to the coupling with the metal M1, the dissolution rate increased from the value $j_{(M1)}$ to the value of the couple. This increase in speed is necessary to compensate for the decrease in the corrosion rate of the metal M1.

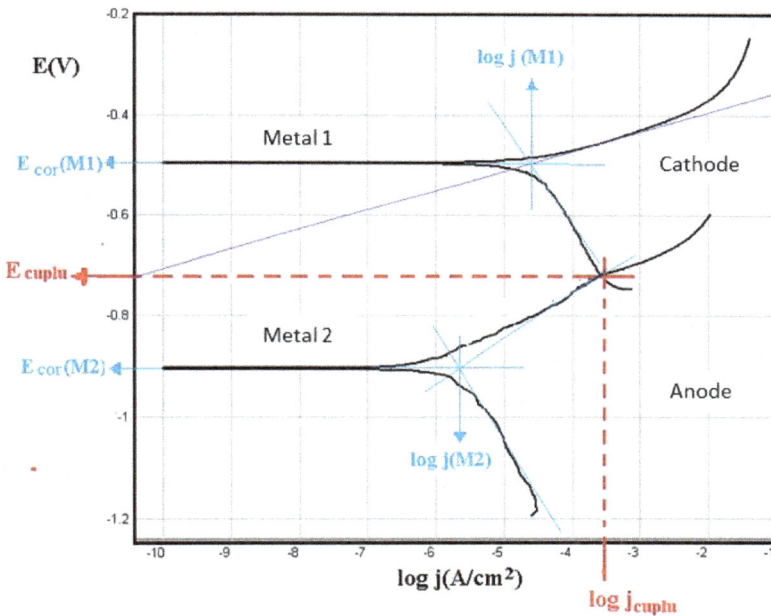

Figure 2.9.   Impedance vector in Cartesian coordinates

The coupling of two different metals can thus be used to protect the material from corrosion by coupling that metal with a less noble metal that corrodes. The process is called cathodic protection or protection with sacrificial anodes.

Thus, sacrificial anodes for steel protection are currently made of three metals: zinc, magnesium and aluminium. Zinc has a potential relative to the steel of around 0.25 V in, a value negative enough to ensure the cathodic protection of steel in environments with not too high resistance. Magnesium, due to its more electronegative potential, ensures the protection of steel even in soils with high resistivity. Aluminium has advantages for use as protective anodes against Mg and Zn (a rather negative potential and a lower electrochemical equivalent), but the pronounced tendency towards passivation of aluminium limits the field of applicability [58].

Experience has shown that increasing the ratio between the cathode area and the anode area increases the dissolution rate of the anode and decreases the corrosion rate of the cathode. It can therefore be admitted that the speed of a galvanic corrosion process of a less noble metallic material, when coupled with a nobler material, is all the greater as the surface of the nobler metal or alloy is greater than that of the metal. or less noble alloy.

References

[1]   Tahamtan, S.; Fadavi Boostani, A. Quantitative Analysis of Pitting Corrosion Behavior of Thixoformed A356 Alloy in Chloride Medium Using Electrochemical Techniques. *Materials & Design* **2009**, *30*, 2483–2489. https://doi.org/10.1016/j.matdes.2008.10.003

[2]   Dwivedi, D.; Lepková, K.; Becker, T. Carbon Steel Corrosion: A Review of Key Surface Properties and Characterization Methods. *RSC Advances* **2017**, *7*, 4580–4610. https://doi.org/10.1039/C6RA25094G

[3]   Luca, C.; Stratula-Vahnoveanu, B. *Chimie Generală*; Editura „Gheorghe Asachi": Iasi, 2003; ISBN 973-621-033-2.

[4]   Udrescu, L. *Tratamente de Suprafata Si Acoperiri*; Politehnica: Timisoara, 2000; ISBN 9739389740.

[5]   Urdas, V. Tratamente Termice, Termochimice, Coroziunea Metalelor Si Acoperiri de Suprafata; Univ: Sibiu, 2001; ISBN 9736512991.

[6]   Pajonk, G.; Steffens, H.-D. Corrosion Behaviour of Coated Materials. *Fresenius' Journal of Analytical Chemistry 1997 358:1* **1997**, *358*, 285–290. https://doi.org/10.1007/s002160050408

[7]   Prohaska, M.; Kanduth, H.; Mori, G.; Grill, R.; Tischler, G. On the Substitution of Conventional Corrosion Tests by an Electrochemical Potentiokinetic Reactivation Test. *Corrosion Science* **2010**, *52*, 1582–1592. https://doi.org/10.1016/j.corsci.2010.01.017

[8]   Otero, E.; Pardo, A.; Sáenz, E.; Utrilla, V.; Pérez, F.J. Intergranular Corrosion Behaviour of a New Austenitic Stainless Steel Low in Nickel. *Canadian Metallurgical Quarterly* **1995**, *34*, 135–141. https://doi.org/10.1016/0008-4433(94)00024-E

[9]   Lee, H.T.; Kuo, T.Y. Effects of Niobium on Microstructure, Mechanical
       Properties, and Corrosion Behaviour in Weldments of Alloy 690.
       http://dx.doi.org/10.1179/136217199101537752 **2013**, *4*, 246–256.
       https://doi.org/10.1179/136217199101537752

[10]   Iacoviello, F.; Casari, F.; Gialanella, S. Effect of "475 °C Embrittlement" on
       Duplex Stainless Steels Localized Corrosion Resistance. *Corrosion Science* **2005**,
       47, 909–922. https://doi.org/10.1016/j.corsci.2004.06.012

[11]   Pedeferri (Deceased), P. Stress Corrosion Cracking and Corrosion-Fatigue. **2018**,
       243–273. https://doi.org/10.1007/978-3-319-97625-9_13

[12]   Wipf, D.O. Initiation and Study of Localized Corrosion by Scanning
       Electrochemical Microscopy. *Colloids and Surfaces A: Physicochemical and
       Engineering Aspects* **1994**, *93*, 251–261. https://doi.org/10.1016/0927-
       7757(94)02872-9

[13]   Overman, R.F. Using Radioactive Tracers to Study Chloride Stress Corrosion
       Cracking of Stainless Steels. *Corrosion* **1966**, *22*, 48–52.
       https://doi.org/10.5006/0010-9312-22.2.48

[14]   Ai, Z.; Sun, W.; Jiang, J.; Song, D.; Ma, H.; Zhang, J.; Wang, D. Passivation
       Characteristics of Alloy Corrosion-Resistant Steel Cr10Mo1 in Simulating
       Concrete Pore Solutions: Combination Effects of PH and Chloride. *Materials
       2016, Vol. 9, Page 749* **2016**. https://doi.org/10.3390/ma9090749

[15]   Santos-Leal, E.; Lopez, R.J. Simultaneous Measurement of Acoustic Emission and
       Electrical Resistance Variation in Stress-Corrosion Cracking. *Measurement
       Science and Technology* **1995**, *6*, 188. https://doi.org/10.1088/0957-0233/6/2/010

[16]   Lv, J.; Luo, H. Effects of Strain and Strain-Induced A'-Martensite on Passive
       Films in AISI 304 Austenitic Stainless Steel. *Materials Science and Engineering:
       C* **2014**, *34*, 484–490. https://doi.org/10.1016/j.msec.2013.10.003

[17]   Bejinariu, C.; Burduhos-Nergis, D.-P.; Cimpoesu, N. Immersion Behavior of
       Carbon Steel, Phosphate Carbon Steel and Phosphate and Painted Carbon Steel in
       Saltwater. *Materials* **2021**, *14*. https://doi.org/10.3390/ma14010188

[18]   Burduhos-Nergis, D.-P.; Vizureanu, P.; Sandu, A.V.; Bejinariu, C. Phosphate
       Surface Treatment for Improving the Corrosion Resistance of the C45 Carbon
       Steel Used in Carabiners Manufacturing. *Materials* **2020**, *13*, 3410.
       https://doi.org/10.3390/ma13153410

[19]   Burduhos-Nergis, D.P.; Vizureanu, P.; Sandu, A.V.; Bejinariu, C. Evaluation of the Corrosion Resistance of Phosphate Coatings Deposited on the Surface of the Carbon Steel Used for Carabiners Manufacturing. *Applied Sciences (Switzerland)* **2020**, *10*. https://doi.org/10.3390/app10082753

[20]   Burduhos-Nergis, D.P.; Bejinariu, C.; Sandu, A.V. *Phosphate Coatings Suitable for Personal Protective Equipment*; Materials Research Forum LLC: Millersville, 2021; Vol. 89; ISBN 9781644901113.

[21]   What Is Non-Destructive Testing (NDT)? Methods and Definition - TWI Available online: https://www.twi-global.com/technical-knowledge/faqs/what-is-non-destructive-testing (accessed on 21 September 2021).

[22]   Frankel, G.S. Pitting Corrosion of Metals: A Review of the Critical Factors. *Journal of The Electrochemical Society* **1998**, *145*, 2186. https://doi.org/10.1149/1.1838615

[23]   Novak, P.; Stefec, R.; Franz, F. Testing the Susceptibility of Stainless Steel to Intergranular Corrosion by a Reactivation Method. *Corrosion* **1975**, *31*, 344–347. https://doi.org/10.5006/0010-9312-31.10.344

[24]   Briant, C.L.; Hall, E.L. A Comparison between Grain Boundary Chromium Depletion in Austenitic Stainless Steel and Corrosion in the Modified Strauss Test. *Corrosion* **1986**, *42*, 522–531. https://doi.org/10.5006/1.3583061

[25]   Alvarez, R.B.; Martin, H.J.; Horstemeyer, M.F.; Chandler, M.Q.; Williams, N.; Wang, P.T.; Ruiz, A. Corrosion Relationships as a Function of Time and Surface Roughness on a Structural AE44 Magnesium Alloy. *Corrosion Science* **2010**, *52*, 1635–1648. https://doi.org/10.1016/j.corsci.2010.01.018

[26]   Gordon, R.; Pincheira, J.A. Influence of Surface Condition on the Inspection of Steel Bridge Elements Using the Time-of-Flight Diffraction Method. *Journal of Bridge Engineering* **2010**, *15*, 661–670. https://doi.org/10.1061/(ASCE)BE.1943-5592.0000100

[27]   van der Ent, J.; van Nisselroij, J.; Kopp, F.; Otter, A.; Weli, N.; Judd, S. Automatic Ultrasonic Inspection of Steel Catenary Risers with a Corrosion-Resistant Alloy Layer. **2006**. https://doi.org/10.4043/17898-MS

[28]   Gordon, R.; Pincheira, J.A. Influence of Surface Condition on the Inspection of Steel Bridge Elements Using the Time-of-Flight Diffraction Method. *Journal of Bridge Engineering* **2010**, *15*, 661–670. https://doi.org/10.1061/(ASCE)BE.1943-5592.0000100

[29] Ambrose, J.R.; Kruger, J. Tribo-Ellipsometry: A New Technique to Study the Relationship of Repassivation Kinetics to Stress Corrosion. *Corrosion* **1972**, *28*, 30–35. https://doi.org/10.5006/0010-9312-28.1.30

[30] Ngo, D.; Liu, H.; Sheth, N.; Lopez-Hallman, R.; Podraza, N.J.; Collin, M.; Gin, S.; Kim, S.H. Spectroscopic Ellipsometry Study of Thickness and Porosity of the Alteration Layer Formed on International Simple Glass Surface in Aqueous Corrosion Conditions. *npj Materials Degradation 2018 2:1* **2018**, *2*, 1–9. https://doi.org/10.1038/s41529-018-0040-7

[31] Poksinski, M.; Dzuho, H.; Arwin, H. Copper Corrosion Monitoring with Total Internal Reflection Ellipsometry. *Journal of The Electrochemical Society* **2003**, *150*, B536. https://doi.org/10.1149/1.1618224

[32] Kear, G.; Barker, B.D.; Walsh, F.C. Electrochemical Corrosion of Unalloyed Copper in Chloride Media—a Critical Review. *Corrosion Science* **2004**, *46*, 109–135. https://doi.org/10.1016/S0010-938X(02)00257-3

[33] Perju, M.C.; Savin, C.; Nejneru, C.; Axinte, M.; Achiței, D.C.; Bejinariu, C. Aspects Regarding Instantaneous Corrosion of Nodular Iron in Household Wastewater. *IOP Conference Series: Materials Science and Engineering* **2018**, *374*, 012016. https://doi.org/10.1088/1757-899X/374/1/012016

[34] de Freitas Cunha Lins, V.; de Andrade Reis, G.F.; de Araujo, C.R.; Matencio, T. Electrochemical Impedance Spectroscopy and Linear Polarization Applied to Evaluation of Porosity of Phosphate Conversion Coatings on Electrogalvanized Steels. *Applied Surface Science* **2006**, *253*, 2875–2884. https://doi.org/10.1016/j.apsusc.2006.06.030

[35] Dexter, S.C.; Gao, G.Y. Effect of Seawater Biofilms on Corrosion Potential and Oxygen Reduction of Stainless Steel. *Corrosion* **1988**, *44*, 717–723. https://doi.org/10.5006/1.3584936

[36] Yamashita, M.; Nagano, H.; Oriani, R.A. Dependence of Corrosion Potential and Corrosion Rate of a Low-Alloy Steel upon Depth of Aqueous Solution. *Corrosion Science* **1998**, *40*, 1447–1453. https://doi.org/10.1016/S0010-938X(98)00041-9

[37] Benea, L.; Simionescu, N.; Mardare, L. The Effect of Polymeric Protective Layers and the Immersion Time on the Corrosion Behavior of Naval Steel in Natural Seawater. *Journal of Materials Research and Technology* **2020**, *9*, 13174–13184. https://doi.org/10.1016/j.jmrt.2020.09.059

[38] Oprisan, B.; Vasincu, D.; Lupescu, S.; Munteanu, C.; Istrate, B.; Popescu, D.; Condratovici, C.P.; Dimofte, A.R.; Earar, K. Electrochemical Analysis of Some Biodegradable Mg-Ca-Mn Alloys. *REV.CHIM.(Bucharest)* ♦ **2019**, *70*, 4525. https://doi.org/10.37358/RC.19.9.7565

[39] Stern, M. The Mechanism of Passivating-Type Inhibitors. *Journal of the Electrochemical Society* **1958**, *105*, 638–647. https://doi.org/10.1149/1.2428683

[40] ASTM F746 - 04(2021) Standard Test Method for Pitting or Crevice Corrosion of Metallic Surgical Implant Materials. *Annual Book of ASTM Standards* 2021.

[41] Kelly, R.G.; Scully, J.R.; Shoesmith, D.W.; Buchheit, R.G. Electrochemical Techniques in Corrosion Science and Engineering. *Marcel Dekker Inc, New York, Basel.* **2003**. https://doi.org/10.4236/aasoci.2012.24032

[42] Corrosion Mechanisms in Theory and Practice. *Corrosion Mechanisms in Theory and Practice* **2002**. https://doi.org/10.1201/9780203909188/corrosion-mechanisms-theory-practice-philippe-marcus

[43] Poursaee, A. Determining the Appropriate Scan Rate to Perform Cyclic Polarization Test on the Steel Bars in Concrete. *Electrochimica Acta* **2010**, *55*, 1200–1206. https://doi.org/10.1016/j.electacta.2009.10.004

[44] Baboian, R.; Haynes, G. Cyclic Polarization Measurements—Experimental Procedure and Evaluation of Test Data. *Electrochemical Corrosion Testing* **1981**, 274-274–9. https://doi.org/10.1520/STP28038S

[45] Bou-Saleh, Z.; Shahryari, A.; Omanovic, S. Enhancement of Corrosion Resistance of a Biomedical Grade 316LVM Stainless Steel by Potentiodynamic Cyclic Polarization. *Thin Solid Films* **2007**, *515*, 4727–4737. https://doi.org/10.1016/j.tsf.2006.11.054

[46] Li, L.; Sagüés, A.A. Chloride Corrosion Threshold of Reinforcing Steel in Alkaline Solutions—Cyclic Polarization Behavior. *Corrosion* **2002**, *58*, 305–316. https://doi.org/10.5006/1.3287678

[47] Castro, E.B.; Vilche, J.R. Investigation of Passive Layers on Iron and Iron-Chromium Alloys by Electrochemical Impedance Spectroscopy. *Electrochimica Acta* **1993**, *38*, 1567–1572. https://doi.org/10.1016/0013-4686(93)80291-7

[48] Cano, E.; Lafuente, D.; Bastidas, D.M. Use of EIS for the Evaluation of the Protective Properties of Coatings for Metallic Cultural Heritage: A Review. *Journal of Solid State Electrochemistry* **2010**, *14*, 381–391. https://doi.org/10.1007/s10008-009-0902-6

[49] Jegdić, B.V.; Bajat, J.B.; Popić, J.P.; Stevanović, S.I.; Mišković-Stanković, V.B. The EIS Investigation of Powder Polyester Coatings on Phosphated Low Carbon Steel: The Effect of NaNO2 in the Phosphating Bath. *Corrosion Science* **2011**, *9*, 2872–2880. https://doi.org/10.1016/j.corsci.2011.05.019

[50] M, I.; A, T.; K, W.; K, N. Deviations of Capacitive and Inductive Loops in the Electrochemical Impedance of a Dissolving Iron Electrode. *Analytical sciences : the international journal of the Japan Society for Analytical Chemistry* **2002**, *18*, 641–644. https://doi.org/10.2116/analsci.18.641

[51] Fattah-alhosseini, A.; Taheri Shoja, S.; Heydari Zebardast, B.; Mohamadian Samim, P. An Electrochemical Impedance Spectroscopic Study of the Passive State on AISI 304 Stainless Steel. *International Journal of Electrochemistry* **2011**, *2011*, 1–8. https://doi.org/10.4061/2011/152143

[52] Tang, N.; van Ooij, W.J.; Górecki, G. Comparative EIS Study of Pretreatment Performance in Coated Metals. *Progress in Organic Coatings* **1997**, *30*, 255–263. https://doi.org/10.1016/S0300-9440(96)00691-1

[53] Bastos, A.C.; Ferreira, M.G.; Simões, A.M. Corrosion Inhibition by Chromate and Phosphate Extracts for Iron Substrates Studied by EIS and SVET. *Corrosion Science* **2006**, *48*, 1500–1512. https://doi.org/10.1016/j.corsci.2005.05.021

[54] Brug, G.J.; van den Eeden, A.L.G.; Sluyters-Rehbach, M.; Sluyters, J.H. The Analysis of Electrode Impedances Complicated by the Presence of a Constant Phase Element. *Journal of Electroanalytical Chemistry and Interfacial Electrochemistry* **1984**, *176*, 275–295. https://doi.org/10.1016/S0022-0728(84)80324-1

[55] Chen, M.; Du, C.Y.; Yin, G.P.; Shi, P.F.; Zhao, T.S. Numerical Analysis of the Electrochemical Impedance Spectra of the Cathode of Direct Methanol Fuel Cells. *International Journal of Hydrogen Energy* **2009**, *34*, 1522–1530. https://doi.org/10.1016/j.ijhydene.2008.11.072

[56] Hack, H.P. Galvanic Corrosion. *Reference Module in Materials Science and Materials Engineering* **2016**. https://doi.org/10.1016/B978-0-12-803581-8.01594-0

[57] Burduhos Nergis, D.P.; Nejneru, C.; Burduhos Nergis, D.D.; Savin, C.; Sandu, A.V.; Toma, S.L.; Bejinariu, C. The Galvanic Corrosion Behavior of Phosphated Carbon Steel Used at Carabiners Manufacturing. *Revista de Chimie* **2019**, *70*, 215–219. https://doi.org/10.37358/RC.19.1.6885

[58] Atrens, A.; Shi, Z.; Song, G.L. Numerical Modelling of Galvanic Corrosion of Magnesium (Mg) Alloys. *Corrosion of Magnesium Alloys* **2011**, 455–483. https://doi.org/10.1533/9780857091413.3.455

[59] Vargel, C. Galvanic Corrosion. *Corrosion of Aluminium* **2020**, 295–315. https://doi.org/10.1016/B978-0-08-099925-8.00025-9

Chapter 3

Methods of Corrosion Protection by Depositing Metallic Protective Layers

3.1. Introduction

The choice of anticorrosive protection means is made both based on the kinetic knowledge of the anodic and cathodic reactions that contribute to the determination of the behaviour of metals and alloys and on the base of the knowledge of the parameters that influence their behaviour.

The covering processes consist of the deposition on the material surfaces of some metallic or non-metallic protective layers.

The protective layers can be inorganic materials (metals, enamels and ceramics) or organic materials (plastics, varnishes or paints) which are obtained by the following methods: chemical, electrochemical and thermomechanical [1].

Coatings can perform the following functions [2,3]:

- chemical and biological protection;
- gives some physical-mechanical properties such as insulators, radiation damping, refractoriness, surface compression stresses, high hardness, low coefficient of friction etc.;
- sanitary and decorative;
- informative: signaling, coding or camouflage.
- To achieve coatings with high durability one must consider the following factors:
- - the condition of the base material surface (clean and with a certain roughness);
- - the properties of the base material;
- - the thickness of the coating;
- - the chemical composition and the properties of coating products;
- - the coating method.

If a certain factor is not applied properly, the quality of the protective layer is adversely affected.

To obtain a durable protective coating, the impurities (oxides, greases, oils, metallic or non-metallic inclusions etc.) must be removed by degreasing, pickling or by mechanical operations. The mechanical method also achieves the roughness required by the technological coating process [4,5].

A good quality protective layer must be: continuous, free of pores, chemically stable, mechanically resistant during operation and have a uniform thickness.

3.2. Preparation of metal surfaces for coating

To achieve the condition of adhesion to the surface, the metal surfaces are prepared by mechanical operations (mechanical processing), chemical and electrochemical methods (degreasing and pickling) [6], according to Fig. 3.1.

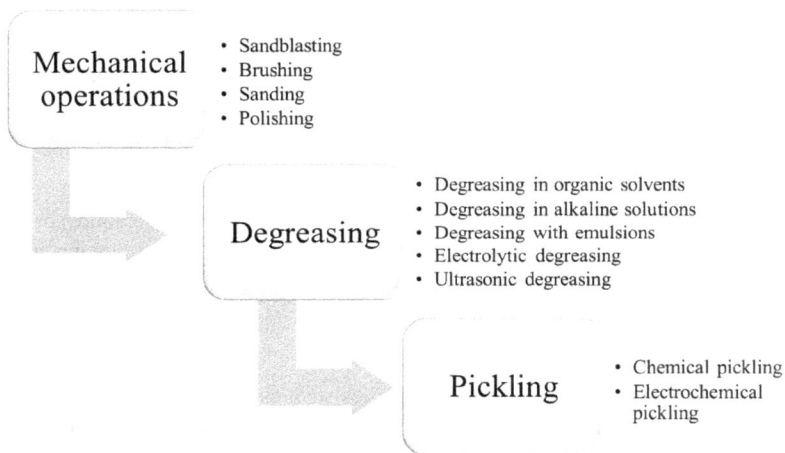

Mechanical operations
- Sandblasting
- Brushing
- Sanding
- Polishing

Degreasing
- Degreasing in organic solvents
- Degreasing in alkaline solutions
- Degreasing with emulsions
- Electrolytic degreasing
- Ultrasonic degreasing

Pickling
- Chemical pickling
- Electrochemical pickling

Figure 3.1.   Stages of surfaces preparation for further coating.

Mechanical operations for cleaning the inclusions of oxides and roughness on surfaces are performed by: sandblasting, brushing, sanding and polishing [7].

Sandblasting is performed on the cast or forged parts with the help of the abrasive action of sand or cast iron particles projected on the metal surface with compressed air. Also, by sandblasting (sand, carborundum, electro-corundum) surfaces with higher roughness are obtained. Also, the sandblasting operation leads to the increase of the fatigue resistance of the substrate, due to the compression stresses generated by the surface hardening [8]. After sandblasting, the surfaces are cleaned to remove dust and material particles that have been sandblasted.

Brushing is done with wire brushes made of animal hair or plastic, by moving them manually or mechanically by rotation [9].

Materials Research Forum LLC
https://doi.org/10.21741/9781644901670

Sanding is done by removing a very thin layer on the surface with the help of abrasive materials of different grains. Abrasive paper or electro-corundated disc is used for sanding [10].

Polishing is achieved by a superior finishing of the surfaces with the help of polishing pastes [11]. For polishing it can be used: cloth soaked in chromium oxide, Vienna lime, Al_2O_3 (metallographic alumina) or silicon carbide (SiC).

Sanding and polishing can be done by vibration.

Degreasing is the operation of removing the grease by dissolving, emulsifying and saponifying. It can be made in organic solvents (hydrocarbons, alcohols, acetone), in alkaline solutions, by wiping, immersion, jet or vapour [12].

Degreasing in organic solvents consists in dissolving saponifiable and unsaponifiable fats. The organic solvents are classified into two categories:

- Flammable - gasoline, kerosene, benzene or toluene;
- Non-flammable - carbon tetrachloride, perchlorate-ethylene, trichlor-ethylene etc.

Due to their tendency to ignite at relatively low temperatures, flammable organic solvents are not widely used. Non-flammable organic solvents allow degreasing at high temperatures, dissolve oils and fats very well and do not react much with metals [13,14].

The most suitable organic solvents used for degreasing are carbon tetrachloride and perchloroethylene. The disadvantage of these chlorinated solvents is their high toxicity and high cost. This type of degreasing is done by immersion or spraying. After degreasing, the parts are washed and dried.

Degreasing in alkaline solutions is done by immersion in baths or spraying with various hydroxides, carbonates, phosphates and detergents [15].

Alkaline chemical degreasing solutions contain substances that form soluble compounds with saponifiable fats and oils, reduce the cohesive force between the grease film or mechanical impurities and metal, facilitate washing and prevent corrosion of the metal. Such components are sodium and potassium hydroxides, soda ash, phosphates, silicates and surfactant additives[16,17].

Alkaline solutions used for degreasing can be of several types:

- weak alkaline solutions, with pH = 10 ÷ 11 for weak degreasing;
- solutions with medium alkalinity, with pH = 11 ÷ 12 indicated for degreasing before galvanizing;
- strong alkaline solutions, with pH = 12 ÷ 14 for strong degreasing.

Materials Research Forum LLC
https://doi.org/10.21741/9781644901670

The temperature at which degreasing takes place is between $60 \div 90$ °C. In general, the chemical composition of degreasing baths contains sodium hydroxide (NaOH), sodium carbonate (Na_2CO_3) and sodium triphosphate (Na_3PO_4).

Degreasing with emulsions is based on the use of a mixture of organic solvent, emulsifiers and water. The composition of the emulsions is solvent (e.g. oil), emulsifier (e.g. sodium silicate and soap), stabilizer and water [18].

Electrolytic degreasing is performed in electrolytic baths in which the base metal can be anode or cathode. This does not apply to steel parts because it oxidises. Electrolytic degreasing is performed in alkaline solutions, at the cathode or anode, but is usually done at the cathode or combined [19,20].

This method is much more effective compared to the others, being performed at temperatures between $50 \div 80$ °C. The electrolyte contains the same components as for degreasing in alkaline solutions: NaOH, Na_2CO_3, Na_3PO_4, NaCN, KCN. The electrodes used are made of nickel-plated steel, graphite or nickel plates.

Ultrasonic degreasing is applied to small parts, with a complex configuration, and is performed with chlorinated solvents or water with emulsifiers, through vibrations that alternately produce pressure and vacuum. This ensures a high degree of cleaning of the metal surface in a short time and requires special installations (ultrasonic generators), which have a high cost [21,22].

The temperature at which the ultrasonic degreasing is performed is between $45 \div 55$ °C, its duration being a few minutes. For the part to be properly degreased, the metal surface needs to be placed in the direction of the sonic field, so during degreasing, the part must be rotated [23].

Pickling is performed to remove oxides (rust and dirt) from the surface of metals in solutions of acids, acid salts or anodic solution [12].

When interacting with acids, oxides on the metal surface form water-soluble salts. The choice of pickling process depends on the nature of the metal, the character and thickness of the oxide layer on its surface [24].

In general, pickling technology has the following steps: rough degreasing, washing, pickling, washing, drying and temporary protection of the metal against corrosion [25].

Chemical pickling is done by immersing the parts in steel tanks, lined with acid-resistant materials [26].

Electrochemical pickling is performed as an anodic process with the dissolution of the metal and the removal of oxides [27].

Electrochemical polishing is performed for decorative-protective coatings. The parts are machined and then degreased and stripped [28]. The anodic process is carried out by dissolving and passivating macro-lifts and micro-grooves.

3.3. Electrolytic deposition of protective layers

By applying a protective metal layer on the material surface, the service life is extended, the operating conditions are improved and the consumption of the deficient metal is reduced. Depending on the destination, the metal coatings can be decorative-protective and special (wear-resistant, friction-resistant etc.) [29].

The electrolytic deposition is an electrochemical process by which corrosion protection or improvement of the mechanical properties of some metals is achieved. This consists in introducing the parts into electrolytic baths with metal salts solutions, neutralizing the charges of the hydrated (metallic) ion, and the neutral atom is incorporated in the crystalline network of the metal [30,31].

The base metal is the cathode and the anode of the direct current source can be an inert metal or the metal used for coating which dissolves in solution as ions moves and deposit at the cathode forming the protective layer.

Electrolysis can be done with a soluble cathode, such as chromium plating or with a soluble anode, such as galvanizing, copper plating or nickel plating.

The protection mechanism of the metal coatings is different and depends on the ratio in which the electrochemical equilibrium potential of the deposited metal is found compared to the potential of the base metal (substrate). Depending on this criterion, the metal coatings are anodic or cathodic [31,32].

Anodic coating occurs when the base metal part is attached to the cathode of a direct current source and the metal used for coating the anode. Anodic coatings are those in which the electrochemical equilibrium potential of the deposited metal is more electronegative than that of the base metal. If are cracks, exfoliations, pores in the protective layer, the local galvanic elements can be formed in these discontinuities, in which the deposited metal being more electronegative becomes the anode and corrodes, and the base metal becomes the cathode, being protected [31–33]. The disadvantage of these types of coatings is that, over time, they will be covered with corrosion products, worsening their appearance. Examples of anodic coatings: zinc and cadmium layers deposited on steel, tin deposited on copper etc.

The cathodic coating is achieved if the deposited metal has a more electropositive potential than the base metal, which acts as an anode [34]. The existence of discontinuities in the protective layer causes it to become a cathode, and the base metal to become anode, corroding. To be protective, the cathodic layers must have sufficient thickness and must be continuous, thus avoiding the formation of corrosion micro piles. The reduction of porosity is also obtained by depositing successive layers of different metals (Cu-Ni-Cr).

The electrolysis bath contains a compound of the metal which will be deposited in certain concentrations, buffers - to maintain a constant pH of the solution, inorganic substances - to increase the electrical conductivity, as well as special additives - to improve porosity, adhesion, gloss and the structure of the protective layer [4,5].

The quality of the coating depends on the surface finish and the working parameters: electrolyte composition, current density, bath temperature, stirring the solution etc.

The deposition time depends on the thickness of the layer [4,5], depending on the technical conditions and is calculated with the formula (eq 3.1):

$$\tau = \frac{0.6\delta \cdot \rho}{k \cdot J \cdot \eta} \qquad (3.1)$$

where: δ – the thickness of the metal layer, μm

ρ – the density of the deposited metal, g/cm^3

K- the electrochemical equivalent of the metal, g/Ah

J- the current density, A/dm^2

η – the current efficiency, %

If the protective metal does not adhere to the base metal surface, several successive metals will be deposited. For example: in the case of nickel plating, first will be deposited a copper layer, then a nickel layer and finally a chrome layer. The thickness of the deposited layers is of the order of hundreds of millimetres.

The method of preparing the metal surface before the coating depends on the deposition method chosen, the degree of impurity of the base metal surface and the desired quality of the surface preparation.

An electrolytic deposition is carried out in carbon steel tanks, lined with plastic or rubber, resistant to the chemical action of the electrolyte and working temperatures, or stainless-steel tanks and equipped with heating installations, thermostats, filters, stirrers, ventilation systems etc. Coating of small parts with metal layers is done in rotating electrolytic baths [35].

The most used metals for protective coating are shown in Fig. 3.2. Also, the electrolytic deposition can be done with the metal alloys shown in the figure, such as Cu-Zn, Sn-Pb, Ni-Co etc.

Figure 3.2.   Metals used for electrolytic deposition.

In the case of electrochemically deposited metal coatings, research in the field has been channelled in several directions, including technologies for surface preparation [36–38], electrochemical processes for applying metals [39], quality control of metal coatings, improvement of properties by changing the parameters or chemical composition of the deposited layer [40–43].

Attempts have also been made to replace toxic cyanide electrolytes with harmless ones [44,45]. A special concern in the field is the alloys deposition, which has very special qualities compared to the metal's deposition. These types of coatings are used in nuclear technology, aerospace, electronics, light industry, in the reproduction of art objects of gold, silver, copper etc. [46–49]

3.3.1. Zinc plating

Zinc is a bluish to white metal. Its main use in industry is to protect metals against oxidation. It has a pronounced electronegative character, it is fragile below 100 °C, but above this temperature, up to temperatures of 200 °C it becomes malleable and ductile.

Zinc coatings are used in many fields, to improve the corrosion resistance of metallic materials on which it is deposited [4,5,35], among which are:

- automotive industry - obtaining parts as control panels, chassis parts, headlights, hoods, coolers etc.
- mechanical processing industry: lathes for metals;
- electronics and electrical engineering industry: for the manufacture of various parts of electrical machinery and equipment, such as electrical doses;
- agricultural industry: fertilizer spreaders;
- household industry: washing machines, automatic washers and dryers etc.;
- transport industry (train): wagon roofs.

Zinc coating protects against corrosion in the air with moderate humidity, in air polluted with flue gases or sulfur gases, in contact with drinking water and contact with petroleum products. Basic zinc salts called "white rust" can form on the surface of Zn coatings.

Electrolytic galvanizing is performed in acidic and alkaline electrolytes with cyanides. The process is carried out with soluble anodes of metallic zinc 99.9%, for glossy galvanizing and with purity, 99.8% for matt galvanizing [35].

Cyanide electrolytes ensure a fine crystalline microstructure of the deposition and a uniform thickness of the layer. These, being toxic, can be replaced with galvanized electrolytes. The electrolyte type is chosen depending on the complexity of the part which must be coated and the adhesion of the deposited layer. Also, the thickness of the zinc layers is established depending on the environment in which the parts work (industrial, rural, marine, dry climate etc.) [4,5].

To control the appearance and properties of the coating, small amounts of other metals can be added to the galvanizing bath, such as antimony, cadmium, tin, lead or aluminium.

The steps of the galvanizing process are presented in Fig. 3.3

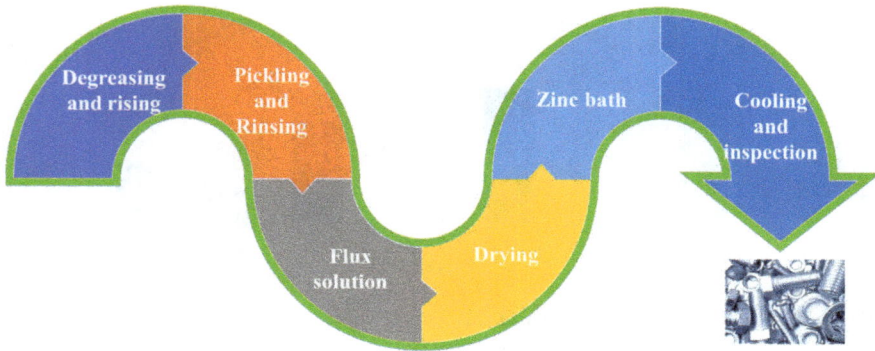

Figure 3.3.   Zinc plating process.

As the base metal surface must be free of impurities, the metal surface must go through the cleaning steps, namely degreasing and pickling.

A satisfactory state of the metal cleaning is maintained by the use of fluxes, so the next step is to introduce the metal into the flux solution.

The fourth stage of the electrolytic deposition process of zinc layers is the immersion of the metal in a zinc bath at high temperatures (depending on the type of electrolyte), thus forming on its surface the zinc layer.

Since zinc coatings are anticorrosive in a maximum humidity of 70%, and under the action of oxygen in the air, CO_2, H_2S and vapours of various substances, basic carbonates are formed, which blacken the layer, the freshly galvanized parts are passivated, acquiring a greenish-yellow colour [35].

After passivation, dehydrogenation is performed at temperatures of 180-200 °C, for 1.5-2 hours.

Literature review

Many researchers have tried over time to improve the chemical and physical properties of zinc layers deposited by electrolytic deposition. For example:

Asgari H. et al. [50]studied how it influences the corrosion resistance and morphology of galvanized zinc coatings, the addition of lead, in different percentages (0.01%, 0.04%, 0.06% and 0.1%), in the zinc bath. It was observed that as the lead content increases, the current density increases and the polarization resistance decreases (Fig. 3.4).

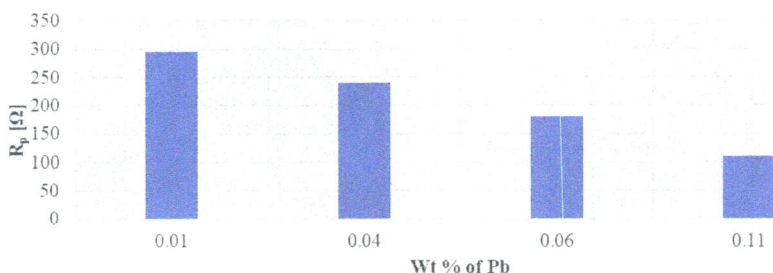

Figure 3.4.   Changing the polarization resistance value according to the lead content in the zinc bath.

Therefore, the corrosion resistance decreases with the increase of the lead content in the zinc bath. Also, with the increase of lead content, the morphology of the samples changes significantly, with an increase in basal planes.

Kania H. et al. [43] studied the corrosion behaviour and microstructure of galvanizing coatings obtained by adding aluminium, nickel and bismuth to the zinc bath. The values for corrosion resistance were compared with those of a pure zinc bath. In terms of the thickness of the deposited layer, it was observed that the zinc layer obtained from the pure zinc solution has a thickness of about 52.5 μm, while the other layer has a thickness of about 51.1 μm. Regarding the morphology of the deposited layers, both have a similar heterogeneous surface, the only difference being the presence of bismuth precipitates on the surface of the layer obtained in the zinc bath with Al, Ni and Bi. Following the electrochemical tests in solution with 3.5% NaCl, the values of current density and corrosion potential were obtained (Fig. 3.5) for the Zn coated sample and for Zn-AlNiBi coated sample.

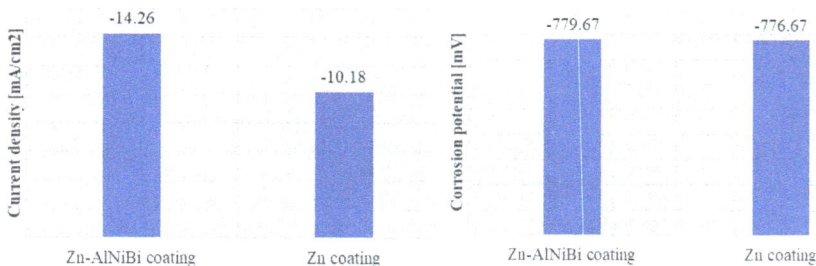

Figure 3.5.   Current density and corrosion potential values for the samples covered in the Zn bath and the Zn-AlNiBi bath.

Thus, it can be seen that the current density value of the zinc-coated sample is lower than that of the Zn-AlNiBi-coated sample, which indicates that the sample coated by immersion in the zinc bath in which aluminium, nickel and bismuth were added has a lower corrosion resistance compared to the Zn-coated sample. This aspect is confirmed by the other two corrosion tests performed: neutral salt spray and a humid atmosphere containing sulfur. The lower corrosion resistance can be explained by the appearance of galvanic torques between bismuth and zinc.

Grandhi S. et al. [51] investigated the corrosion behaviour and morphology of galvanized zinc coating by adding to the zinc bath manganese in percentages of 0.1, 0.5 and 1.2%. To prevent the formation of manganese oxides formed on the surface, another chemical element was added in the bath which has a higher tendency to oxidize, namely aluminium, in a percentage of 2%. It was observed that the addition of manganese in the zinc bath led to the alteration of the morphology of the layer, but without changing the layer adhesion to the steel surface. The corrosion resistance of the obtained samples was determined by linear polarization method in 3.5% NaCl solution, the results being compared with that of a zinc-coated sample (Figs. 3.6. and 3.7). It can be seen that by adding manganese in the zinc bath the current density decreases significantly, while the values of the corrosion potential are approximately the same.

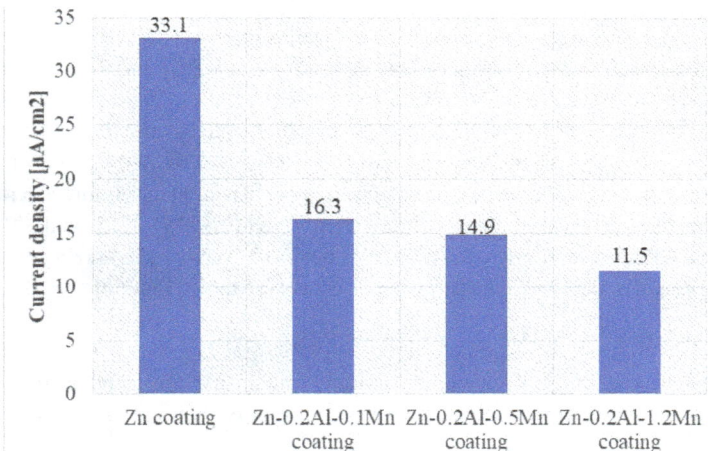

Figure 3.6.   Current density values for the studied samples.

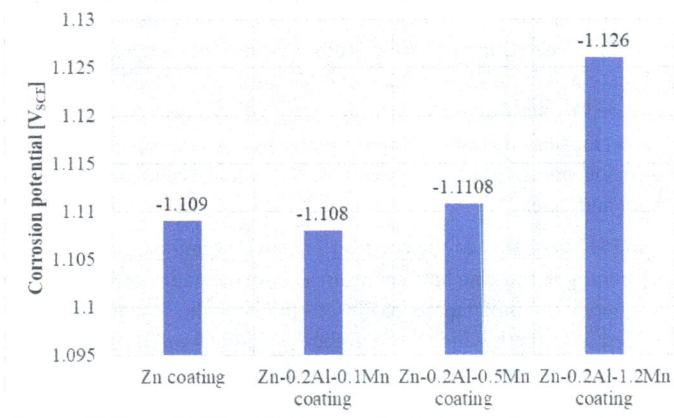

Figure 3.7.   Corrosion potential values for the studied samples.

Analyzing the obtained values of current density and corrosion potentials for zinc-coated and Zn-Mn-Al coated samples, it was observed that the corrosion resistance of the coated sample in the bath containing 1.2% Mn is 65% higher compared to Zn coated sample.

As can be seen, many researchers have tried to obtain superior properties of the coating by changing the chemical composition of zinc baths. For example, Peng S. et al. [52] added aluminium and antimony to the zinc bath and subsequently studied, by various methods (neutral salt spray test, potentiodynamic polarization and electrochemical impedance spectroscopy) the corrosion resistance of the obtained layer. Also, Al-Negheimish A. et al. [53] demonstrated that the corrosion resistance of steel bars used to reinforce concrete can be improved by adding aluminium in different amounts (10%, 15%, 20% and 30%) to the zinc bath. In addition, to improve the properties obtained from the galvanizing process, the impact of pre-galvanizing treatments on the coating was studied [54–56].

3.3.2. Tin plating

Tin is a metal that has high chemical stability, having a high chemical resistance in humidity. It is mainly deposited on parts obtained from copper and its alloys or on steel parts (previously copper-plated or brass-plated).

Tin plating is used in many areas, [5,35] such as:

- in the food industry - for example, when covering plates used in the manufacture of the cans or when covering cutlery and kitchen utensils;

- in the electronics and electrical engineering industry - for covering electrical contacts, copper cables etc.;
- aerospace industry;
- jewelry manufacturing;
- to cover piston rings or aluminium pistons;
- for the protection of surfaces against thermochemical treatments etc.

The tin coating protects against corrosion in several corrosive environments, such as: in the atmosphere, in seawater, in ammonia, in diluted organic acids and liquid fuels. However, it has low corrosion resistance in mineral acids and hydroxides.

The electrolytic deposition of tin can be done with two different types of electrolytes:

- acid electrolytes: with soluble Sn^{2+} salts;
- alkaline electrolytes, in which tin is in the form of Sn^{4+}.

The electrolytes used for tin coating are of three types [35]:

- sulphate (chemical composition: stannous sulphate, gelatin, cresol-sulfonic acid, sulfuric acid, β-naphthol), the coating taking place at a temperature of 20-25 °C;
- fluoroborate (chemical composition: stannous fluoroborate, gelatin, free boric acid, free fluoroboric acid, metallic tin, β-naphthol), the coating taking place at a temperature of 20-35 °C;
- stannate (chemical composition: sodium stannate, sodium perborate, sodium hydroxide, sodium acetate), the coating taking place at a temperature of 60-80 °C.

The adhesion of tin layers is higher in the case of depositions made in alkaline baths compared to those obtained in acidic baths.

The steps of the electrolytic deposition process of the tin layers are presented in Figs. 3.8.

For the surface to be uniformly coated, before the electrolytic deposition of the tin, the part must go through several cleaning steps, depending on the condition of the surface on which the tin layer will be deposited. Usually, the cleaning steps are: grit blasting (to remove impurities from the surface), boiling (in order to remove grease on the surface without using chemical additives), degreasing (in order to remove the grease and oil) and rising (to remove the chemical compounds from the metal surface that remained after degreasing).

The next step after cleaning the metal surface is electroplating. Before the actual plating, the tin plating bath must be prepared, depending on the coating properties we want to obtain and depending on the nature of the base metal.

In order to improve the corrosion resistance of the tin coating or to prevent hydrogen embrittlement, it is necessary a post-treatment stage. Also, after passivation, the surface is greased with a thin layer of oil.

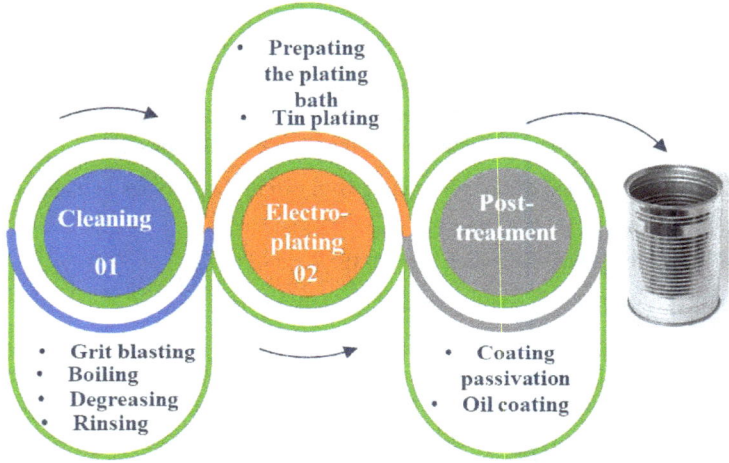

Figure 3.8.   *Flow tin plating process.*

The product obtained from the electrolytic deposition process of tin can be considered a material consisting of five layers (Fig. 3.9).

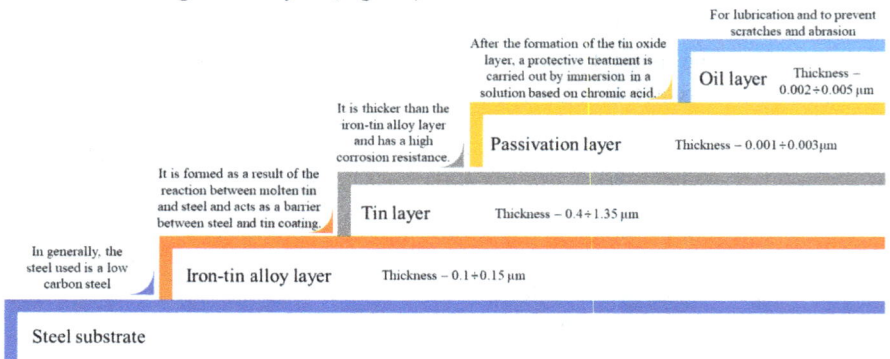

Figure 3.9.   *The structure of tin coating on steel, obtained by electrolytic deposition.*

Literature review

Many researchers have studied the effects of changes in electrolyte deposition parameters on the properties of tin coatings. For example, Sharma A. and Ahn B. [57] studied the effect of bath pH as well as the effect of current density on the properties and morphology of tin coating on copper substrate. The electrolyte used is stannate and the pH of the baths varied from 9 to 14. The samples obtained were studied by XRD, linear polarization method, SEM and micro-hardness. Regarding the corrosion resistance, it was concluded that the tin layers obtained in solutions with pH = 9 or 10 have low corrosion resistance, an aspect also supported by SEM images where it could be seen that the obtained layer is porous. Regarding the current density, it was observed that the thickness of the Sn layer and the coating speed depend on its value. The best properties being obtained at the value of the current density of 15 mA cm^2 and a pH of 13.

Also, Sekar R. et al. [58] studied the effects of adding different additives in the plating bath on the morphology and corrosion resistance properties of the tin coating. The additive added in the plating bath is nil, peptone, gelatin, β-Naphthol, Histidine and polyethene glycol. From the corrosion point of view (linear polarization and electrochemical spectroscopy impedance tests, Figs. 3.10, 3.11 and 3.12), it was observed that the samples obtained in the bath containing peptone have the highest corrosion resistance.

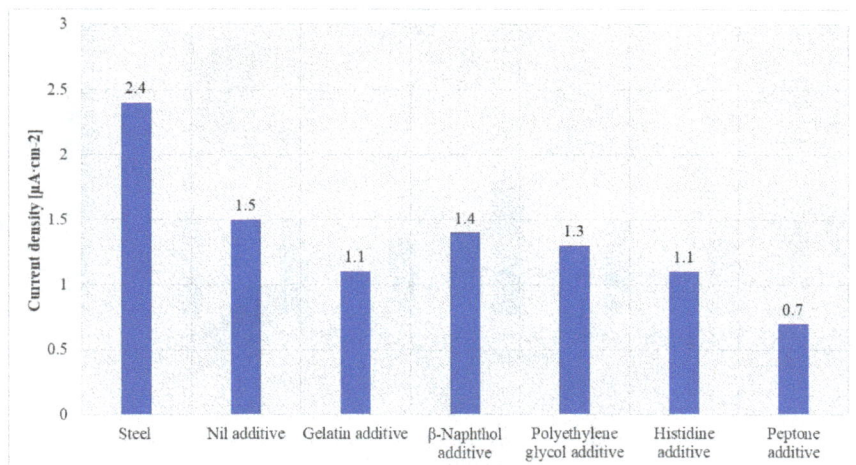

Figure 3.10. The current density values depending on the additive type in the tin plating bath.

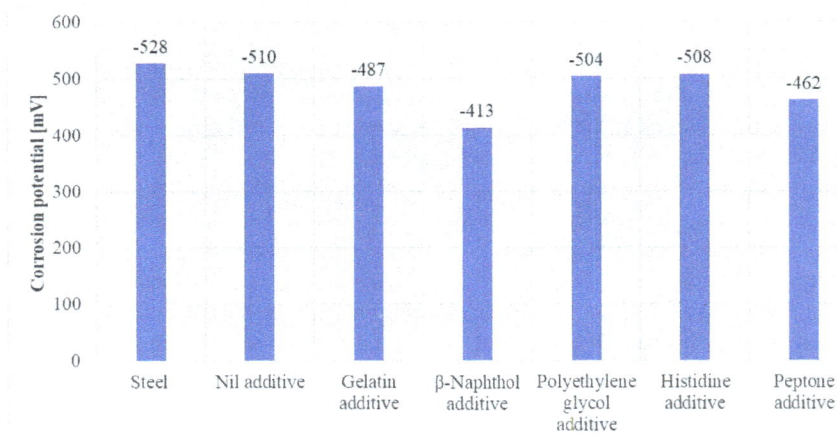

Figure 3.11. The corrosion potential values depending on the additive type in the tin plating bath.

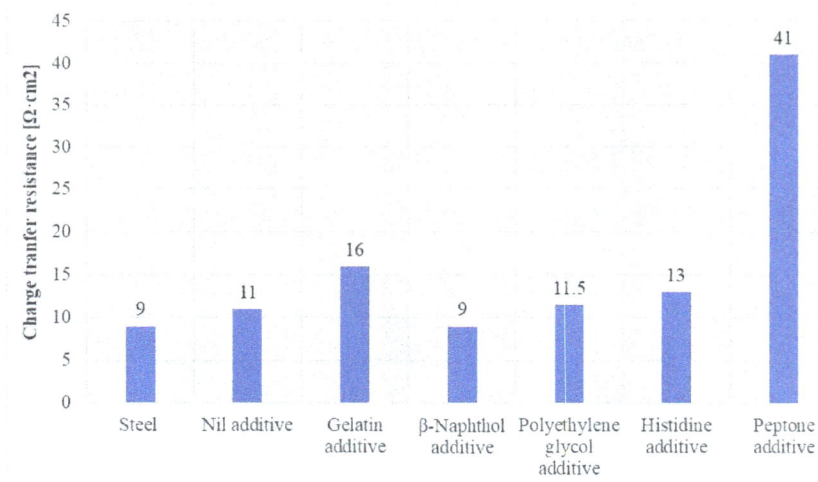

Figure 3.12. The charge transfer resistance values depending on the additive type in the tin plating bath.

A very studied topic in recent years has been the use of tin coatings or coatings with tin alloys on materials used for lithium-ion batteries. E.g. Deng M.J. et al. [59] studied the electrolytic deposition of tin on mesocarbon microbead powder. Also, Li C.F. et al. [60] studied the Sn/LiO_2 coating on stainless steel substrates for lithium batteries.

3.3.3. Chrome plating

Chromium is a white-silver metal that has high corrosion resistance (due to the accentuated tendency of passivation), high resistance to mechanical wear and a pleasant appearance. It is deposited electrochemically (cathodic) on almost all metals (nickel, copper, zinc, aluminium etc.).

The chrome coating not only helps protect against corrosion but also ensures good resistance to mechanical wear. The high stability of chromium coatings against oxidizing media (nitric acid) is due to the formation of a thin and invisible layer of chromium oxide.

Due to its high light reflection capacity, chrome plating can replace silver plating. Because, although silver has a higher coefficient of reflection than chromium, it is not stable in the atmosphere, reacting with hydrogen sulfide and forming a layer of black silver sulfide [4,5].

The chromium layer has a high chemical resistance up to temperatures of 500 °C but is attacked by halogenated acids and sulfuric acid.

Depending on the parameters used for electrolytic deposition of chrome layers, there are three types of chrome plating. The most important and used are the decorative chrome plating and the hard chrome plating, presented in Fig. 3.13.

Another chrome plating type is black chrome plating, generally used in the solar energy industry or the manufacture of thermal and optical appliances. It has multiple properties, such as good thermal conductivity, high thermal resistance (up to 600 °C), good light absorption capacity etc. Black chrome plating can be done directly on steel or an intermediate layer of nickel. Black chromium coatings are obtained under certain conditions, when chromium plating is performed in an electrolyte with a low sulfuric acid content, using a low temperature and a very high current density [35].

Another type of chrome plating is porous chrome plating. Porous chrome coatings are normal chrome coatings, on which numerous very small pores are created by certain methods, due to which the wetting surface of the chrome layer by the oil becomes very good. These can be obtained by three methods:

- mechanical method - consists in roughing the base surface using cutting tools or by sandblasting with sand or cast iron;

- chemical method - consists in processing chromed surfaces in solutions based on sulfuric acid or hydrochloric acid;
- electrochemical method - consists of the additional anodic processing of the chromed part in an electrolyte like the one in which the chrome plating was performed.

Porous chrome plating is used on the cylinder liners of diesel engines, bushings, bearings etc.

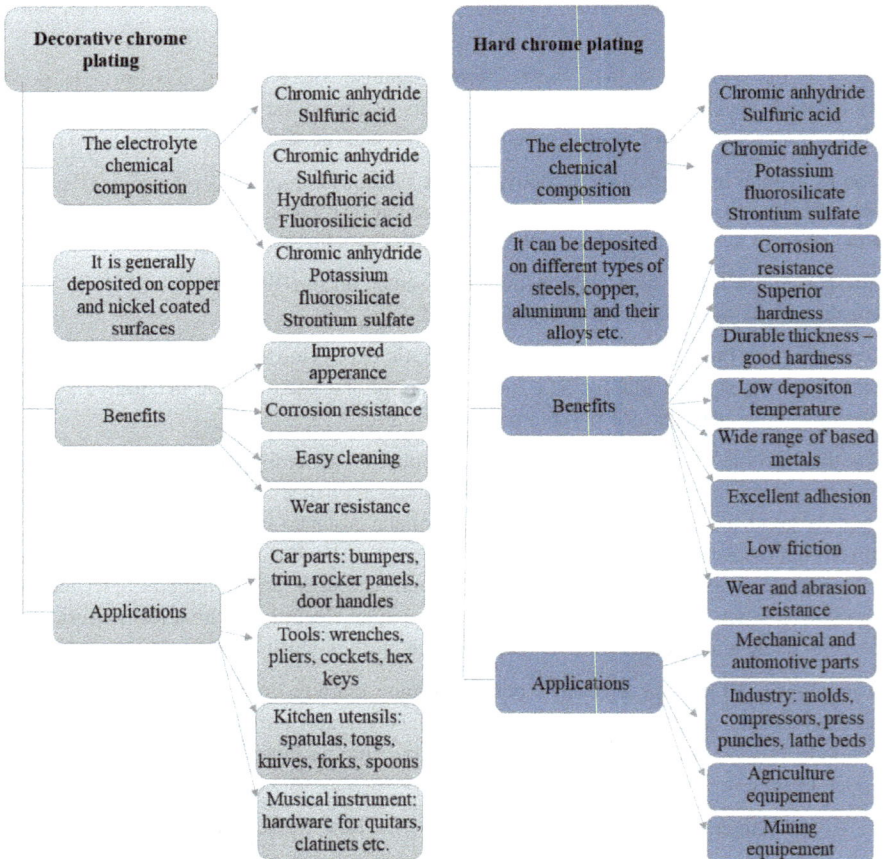

Figure 3.13. Chrome plating types.

The steps of the chrome plating are shown in Fig. 3.14.

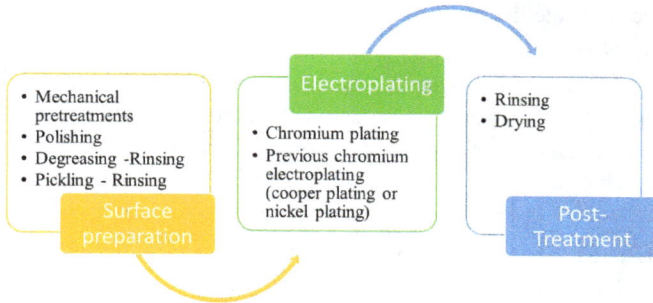

Figure 3.14. Chrome plating process.

The surface preparation steps may be different depending on the surface condition of the base metal and its nature.

Due to the properties of chromium to passivate, it becomes more electropositive compared to iron, so that in the case of galvanic chromium-iron torque, it does not protect iron against corrosion. Therefore, after coating the steel with chrome, a nickel or copper coating is recommended. Also, due to the dissolution reaction of copper with the chrome plating solution, it is not recommended to cover copper or copper alloys.

In the case of decorative chrome plating, if the nickel plating was carried out in an electrolyte without the addition of gloss, the coating obtained is polished by mechanical processes, followed by degreasing in lime, washing in running water and chrome plating. If the nickel plating was performed in an electrolyte with the addition of gloss, the coated part is washed in cold running water and then coated with chrome. After drying, the part is subjected to technical inspection.

Literature review

Over time, the properties of chrome coatings have been studied depending on the process parameters or the thickness of the chromium layer. For example, Santos B.A.F. et al. [61] studied the properties of the chromium layer depending on its thickness (11-12 µm, 13-16 µm, 19-20 µm, 20-22 µm and 29-34 µm). The corrosion resistance of the samples was determined by linear polarization method in solution with 3.5% NaCl and salt spray chamber test, the samples being exposed to a solution with 5% NaCl. Following the salt spray chamber test, it was observed that points or localized corrosion areas appear on the surface, the sample with the highest thickness having the highest corrosion resistance. This

Materials Research Forum LLC
https://doi.org/10.21741/9781644901670

aspect is also confirmed by the linear polarization tests, where it could be observed that with the increase of the thickness of the deposited layer, the corrosion resistance of the sample also increases.

Also, Li J. et al. [62] studied the hardness and corrosion properties of trivalent chromium plating coatings. The parameters (time, current density and tank pressure) used in the plating process are different. From the corrosion resistance point of view, the researcher recommends being added in the plating solution a nanometer additive to improve the corrosion resistance of the trivalent chromium coatings.

Dongwook L. et al. [63] compared the properties of the chromium coatings obtained by pulse-reverse electroplating and those obtained by direct current electroplating, also they studied the effect of anodic time on the properties of the chromium layers. Regarding the corrosion resistance of the samples, it was observed that the best corrosion resistance is presented by the chromium layers obtained by pulse-reverse electroplating compared to those obtained by direct current electroplating.

Other researchers have compared different types of coatings to determine which are more suitable in certain applications. For example, Deepak J.R. et al. [64] compared the chrome, copper, nickel and zinc coatings on the steel surface in terms of morphology, hardness and corrosion resistance, to use them in construction applications. The corrosion resistance of the samples was determined by the linear polarization method and by electrochemical impedance spectroscopy in a 3.5% NaCl solution. The values of corrosion potential, current density and subsequent charge transfer resistance were compared (Figs. 3.15, 3.16 and 3.17).

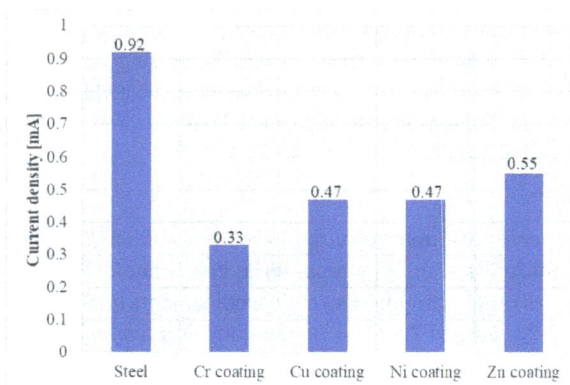

Figure 3.15. The current density values of the steel and Cr, Cu, Ni, Zn coatings.

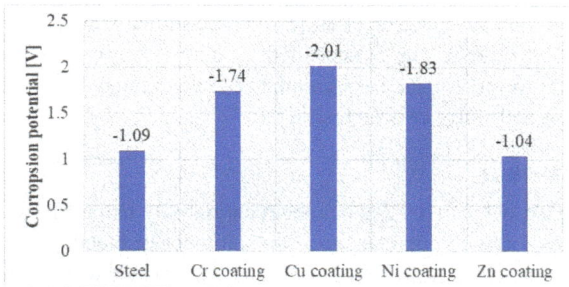

Figure 3.16   The corrosion potential values of the steel and Cr, Cu, Ni, Zn coatings.

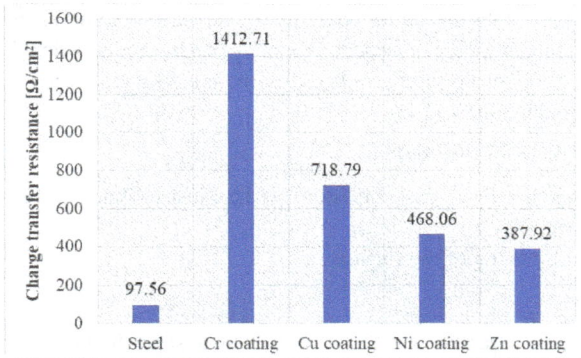

Figure 3.17.   The charge transfer resistance values of the steel and Cr, Cu, Ni, Zn coatings.

These values indicate that the chrome plating has the best corrosion resistance due to the strong passive Cr layer which acts as a barrier between the steel and the corrosive environment.

Viswanathan R. et al. [65] studied the possibility of using chrome coatings on magnesium alloys used in the aerospace industry.

Of course, in search of solutions to improve the properties of materials used in various applications, many researchers have studied the possibility of replacing chrome coatings with other types of coatings. For example, Tay S. L. et al. [66] suggested the replacement of hard chrome plating with cirrus doped electroless nickel coating, due to the excellent properties of corrosion resistance, wear-resistance and hardness.

Other researchers, such as Wang Q. et al. [67] demonstrated that carbide-based coatings namely WC-40Cr3C2-24NiCr layer deposited by high-velocity oxy-fuel (HVOF) are a potential candidate to replace the hard chrome plating coatings due to the highest wear resistance hardness and slurry erosion resistance.

3.3.4. Nickel plating

Nickel is a yellowish-white metal that has chemical stability in the atmosphere. Therefore, it is used in electroplating to obtain a corrosion resistance of metallic materials in contact with atmospheric air, alkaline hydroxides or even in contact with weak organic acids, natural and distilled water. The high stability of nickel is explained by the formation on its surface of a protective film of nickel hydroxide [5]. However, it is strongly attacked, under certain conditions, by chlorine, sulfur compounds and various acids, such as hydrochloric acid, nitric acid, sulfuric acid etc. Depending on the parameters at which the nickel plating is performed (current density, temperature etc.) there are several types of nickel plating, which are presented in Fig. 3.18.

Nickel plating is done for two reasons:

- protection of metals against corrosion: medical instruments, household items, in the food industry etc.
- decorative finishing of the surface

Literature review

Wang Y. et al. [68] studied a method to obtain a super-hydrophobic surface on a magnesium alloy by depositing a nickel layer on the Mg alloy surface. Also, this study was investigated the wettability and corrosion behaviour of the obtained layer. The corrosion resistance was analyzed by linear polarization in 3.5% NaCl solution for three types of samples: Mg alloy sample (1), the Mg alloy sample modified with stearic acid (2) and the Mg alloy sample modified with stearic acid coated with nickel layer (3). The corrosion potential, current density and corrosion rate obtained are presented in Fig. 3.19.

According to the values presented in Fig. 3.19, the sample with the highest corrosion resistance is the Mg alloy sample modified with stearic acid and covered with a nickel layer. This conclusion is supported by the values of the contact angle measured that show that the surface of the third type of sample is super-hydrophobic.

Hiang T. et al. [41] investigated how the current density influences the wettability and corrosion resistance properties of nickel coating. For nickel electroplating was used different values of current density such as 2 A/dm^2, 4 A/dm^2, 6 A/dm^2 and 8 A/dm^2. Also, they studied two types of coatings: nickel coating and nickel coating modified by myristic acid. The corrosion resistance was evaluated by the linear polarization method and EIS. It was observed that depending on the value of the current density the morphology and wettability of the samples are changed. The best mechanical stability and wettability properties were obtained for the 6 A/dm^2 current density. Also, the sample coated with nickel and modified by myristic acid has the highest corrosion resistance.

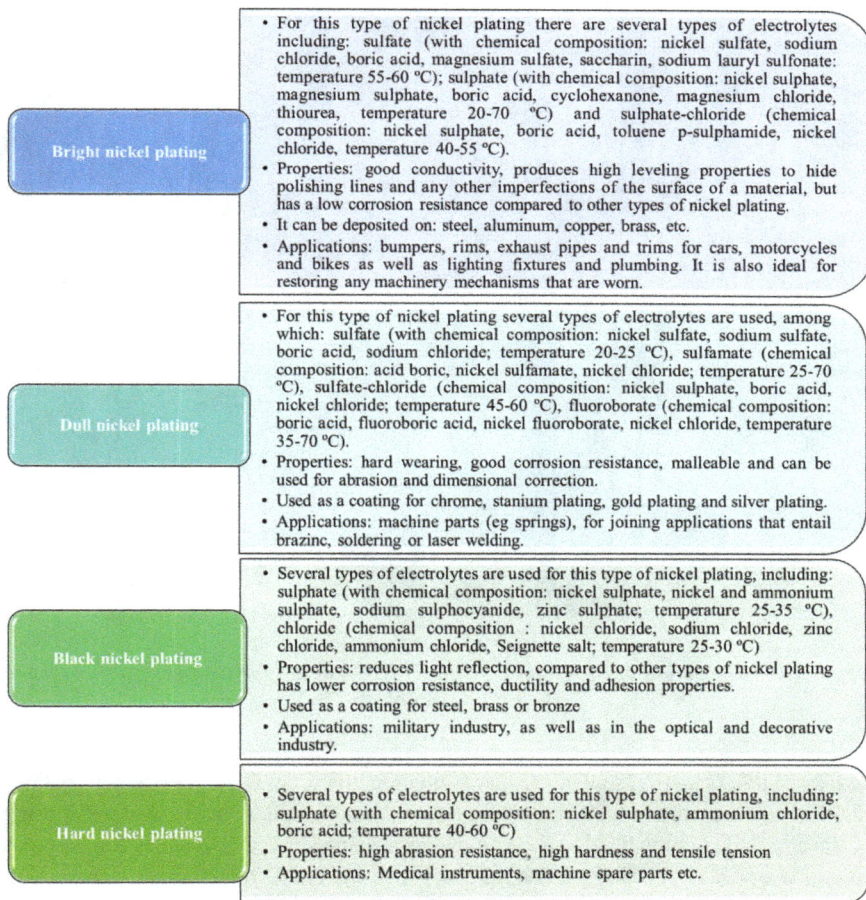

Bright nickel plating

- For this type of nickel plating there are several types of electrolytes including: sulfate (with chemical composition: nickel sulfate, sodium chloride, boric acid, magnesium sulfate, saccharin, sodium lauryl sulfonate: temperature 55-60 ℃); sulphate (with chemical composition: nickel sulphate, magnesium sulphate, boric acid, cyclohexanone, magnesium chloride, thiourea, temperature 20-70 ℃) and sulphate-chloride (chemical composition: nickel sulphate, boric acid, toluene p-sulphamide, nickel chloride, temperature 40-55 ℃).
- Properties: good conductivity, produces high leveling properties to hide polishing lines and any other imperfections of the surface of a material, but has a low corrosion resistance compared to other types of nickel plating.
- It can be deposited on: steel, aluminum, copper, brass, etc.
- Applications: bumpers, rims, exhaust pipes and trims for cars, motorcycles and bikes as well as lighting fixtures and plumbing. It is also ideal for restoring any machinery mechanisms that are worn.

Dull nickel plating

- For this type of nickel plating several types of electrolytes are used, among which: sulfate (with chemical composition: nickel sulfate, sodium sulfate, boric acid, sodium chloride; temperature 20-25 ℃), sulfamate (chemical composition: acid boric, nickel sulfamate, nickel chloride; temperature 25-70 ℃), sulfate-chloride (chemical composition: nickel sulphate, boric acid, nickel chloride; temperature 45-60 ℃), fluoroborate (chemical composition: boric acid, fluoroboric acid, nickel fluoroborate, nickel chloride, temperature 35-70 ℃).
- Properties: hard wearing, good corrosion resistance, malleable and can be used for abrasion and dimensional correction.
- Used as a coating for chrome, stanium plating, gold plating and silver plating.
- Applications: machine parts (eg springs), for joining applications that entail brazinc, soldering or laser welding.

Black nickel plating

- Several types of electrolytes are used for this type of nickel plating, including: sulphate (with chemical composition: nickel sulphate, nickel and ammonium sulphate, sodium sulphocyanide, zinc sulphate; temperature 25-35 ℃), chloride (chemical composition : nickel chloride, sodium chloride, zinc chloride, ammonium chloride, Seignette salt; temperature 25-30 ℃)
- Properties: reduces light reflection, compared to other types of nickel plating has lower corrosion resistance, ductility and adhesion properties.
- Used as a coating for steel, brass or bronze
- Applications: military industry, as well as in the optical and decorative industry.

Hard nickel plating

- Several types of electrolytes are used for this type of nickel plating, including: sulphate (with chemical composition: nickel sulphate, ammonium chloride, boric acid; temperature 40-60 ℃)
- Properties: high abrasion resistance, high hardness and tensile tension
- Applications: Medical instruments, machine spare parts etc.

Figure 3.18. Nickel plating types.

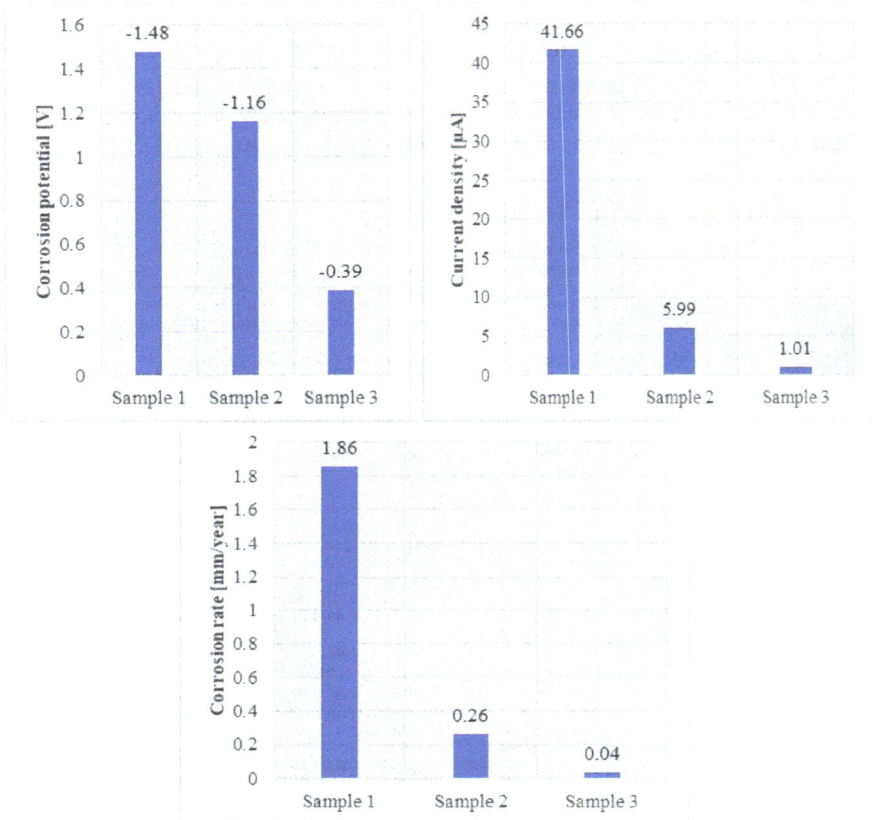

Figure 3.19. The values of corrosion potential, current density and corrosion rate for the studied samples.

Laszczynska A. et al. [69] studied the corrosion properties of nickel-molybdenum coatings with different content of Mo. The corrosion properties of coatings were investigated using the linear polarization method and electrochemical impedance spectroscopy for six types of samples coated with Ni, Ni-Mo (with Mo content of $\cong$ 11.1, 19.4, 21.1, 25.6, 28.4, 31.6). The SEM micrographs (Fig. 3.20) shows a big change in the morphology of the studied layers. The surface of Ni coating is composed of pyramidal grains. With the addition of molybdenum in the plating bath, the surface morphology changes significantly, reaching that at 31.6% Mo in the bath the sample surface to be smooth, not being able to observe the grains.

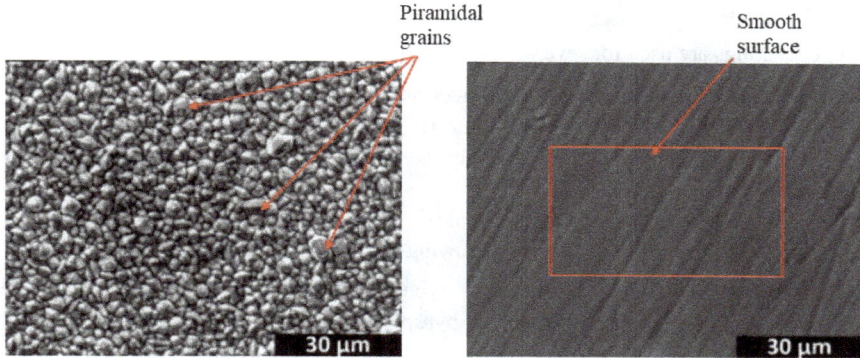

Figure 3.20. The SEM micrographs for (a) Ni coating and (b) Ni-Mo coating with 31.6 Mo content.

From the corrosion point of view for a longer exposure time, it was observed that the samples with the highest corrosion resistance are those that are covered with Ni-Mo (21-28% Mo content), due to the formation of a Mo oxides passive film.

Birlik I. and Ak Azem N.F. [70] studied the influence of the plating bath composition on the morphology and the properties of Ni coatings. Jiang N. et al. [71] investigated a possibility to improve the corrosion resistance of carbon steel in an acidic chloride environment by depositing nickel-phosphorus (Ni-P) / nano-ZnO multilayer coating on its surface. Rahimi E. et al. [72] studied the effect of different contents of H_3BO_3 in the plating bath on the morphology and corrosion resistance of nickel coatings.

3.3.5. Copper plating

Copper is a reddish, easy-to-process metal that has good thermal and electrical conductivity properties.

Due to its low resistance to oxidation in the atmosphere, copper is not used for corrosion protection (especially in the case of steels, it accelerates the corrosion of iron) and not for decorative coatings. However, copper coatings are stable to water and mineral acids, which do not have the action of oxidation [35].

In general, copper coatings are applied before thermochemical treatments to carburize steels and are used as an intermediate layer for nickel plating, chrome plating and silver plating of steels.

Copper coatings are used for several purposes [4,5]:

- the intermediate layer, between steel and other metal, deposited to increase the corrosion resistance and obtain a surface with mirror gloss;
- protection against cementation, because the copper coatings prevent carbon entanglement;
- to protect steel parts against nitriding;
- to reconditioning used parts and drawing or blackmailing operations.

The electrolytes used in copper are of two types: acids (copper sulphate, copper oxalate, etc.) and alkaline (copper cyanide, copper pyrophosphate, etc.).

Considering that one of the main uses of copper coatings is the local protection of the parts against cementation, in Fig. 3.21 the stages of the copper plating process are presented.

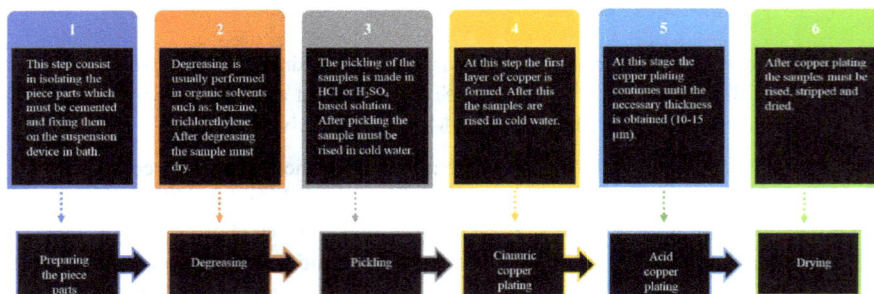

1	2	3	4	5	6
This step consist in isolating the piece parts which must be cemented and fixing them on the suspension device in bath.	Degreasing is usually performed in organic solvents such as: benzine, trichlorethylene. After degreasing the sample must dry.	The pickling of the samples is made in HCl or H₂SO₄ based solution. After pickling the sample must be rised in cold water.	At this step the first layer of copper is formed. After this the samples are rised in cold water.	At this stage the copper plating continues until the necessary thickness is obtained (10-15 µm).	After copper plating the samples must be rised, stripped and dried.
Preparing the piece parts	Degreasing	Pickling	Cianuric copper plating	Acid copper plating	Drying

Figure 3.21. Copper plating process for local protection of parts against cementation.

Literature review

Raghupathy Y. et al. [73] investigated the influence of graphene oxide (GO) on corrosion resistance and morphology of Cu-GO coatings. The corrosion resistance of the studied samples (mild steel sample -MS, steel sample coated with Cu – Cu/MS, steel sample coated with Cu-GO using different baths – Cu-GO1/MS, Cu-GO2/MS and Cu-GO3/MS) was analyzed by linear polarization method in 3.5 % NaCl based solution. The values of corrosion potential and current density obtained are presented in Fig. 3.22.

According to the obtained values, it can be observed that the corrosion resistance of the mild steel is improved by covering its surface with a copper layer. However, we compare the corrosion potential and current density values of all the samples, the Cu-Go samples have the highest corrosion resistance.

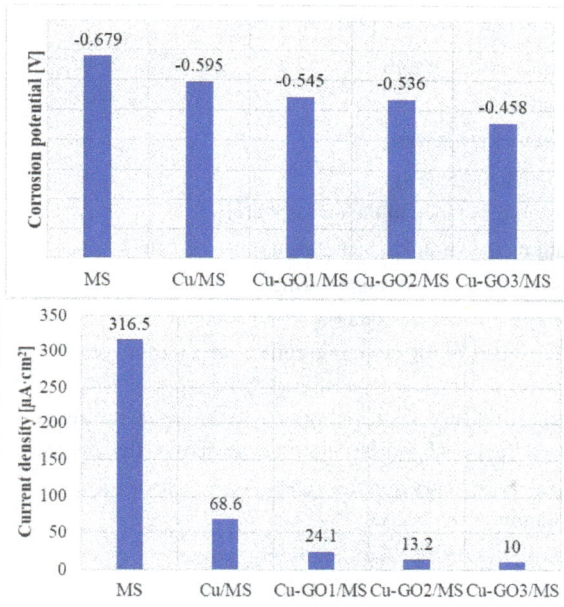

Figure 3.22. The values of corrosion potential and current density for the studied samples.

Many researchers conducted studies related to the use of copper coating on different types of materials. For example, Huang J. et al. [74] studied a method to obtain a high-adhesion Cu layer deposited on polyethene terephthalate, concluding that this method can be used for future flexible printed circuit board industry. Azar G. T. P. et al. [75] analyzed the effect of different CuNP based catalysts on Cu plating on textiles. The study revealed that CuNP catalysts can be used successfully to produce a highly conductive copper coating on textiles. Also, Liu W. [38] investigated the possibility to modify the mica powder surface to be able to be covered with copper. Before Cu plating of mica powder surface, the powder

Materials Research Forum LLC
https://doi.org/10.21741/9781644901670

was activated using nano-nickel and coarsened using ultrasound, obtaining a uniform Cu layer.

3.3.6. Silver plating

Silver is a white metal that has plasticity and malleability properties.

Silver coatings have many benefits [35], such as:

- high light reflective capacity;
- easy to clean/polish;
- good antifriction properties;
- high electrical conductivity;
- high corrosion resistance in the atmosphere.
- Its applications are variants, being used in several fields, such as:
- electronics and electrical engineering - to protect and increase the electrical conductivity of various metals used in the manufacture of electrical appliances;
- decorative industry - for covering cutlery and various art objects, as well as for jewelry;
- in the optical industry, in astronomy - to improve the reflection coefficient of surfaces used in optical and astronomical devices, reflectors etc.;
- in the chemical industry for the protection of chemical apparatus or against the action of alkaline solutions etc.;
- in the industry - to improve the sliding properties of the surfaces subjected to friction.

If high mechanical strength is needed, it is alloyed with antimony, selenium or tellurium. Also, if it is necessary to have a high resistance to wear, a layer of nickel will initially be deposited on the piece on which the silvering is to be made [76,77].

Silver coatings can be made on several metals, such as copper and its alloys, nickel and its alloys, aluminium and its alloys, zinc alloys and steels [78]. In the case of steel parts to improve the adhesion of the silver layer, before silver plating, either copper, brass or nickel plating is made. Highly alkaline electrolytes can be used for silver coatings.

The steps of the silvering process are shown in Figs. 3.23.

Figure 3.23. Silver plating process.

Given that there is a large difference between the value of the electrochemical potential of silver and that of iron, when introducing steel into the silver bath, on its surface occurs the deposition of a silver layer with poor adhesion. Therefore, to prevent the deposition of silver by contact, other metals are initially deposited on the steel parts. To reduce the difference between the two values of the electrochemical potential, the parts can be amalgamated in weak mercury solutions or a preliminary silver plating can be performed in a silver electrolyte with low silver ion content and a high free cyanide content.

3.3.7. Gold plating

Gold electrolytic deposits have wide use in jewelry making, electronics, cosmonautics etc. They have the following properties:

- superior corrosion resistance to other types of galvanic coatings;
- high thermal conductivity;
- high electrical conductivity;
- decorative aspect;
- low wear resistance.

If a high layer hardness is required, nickel-gold alloys or cobalt-gold alloys are used for deposition.

Due to the high price, it is mainly used to cover the parts of special radio-technical and electrotechnical devices, precision devices to protect them against corrosion [79].

The stages of the technological process are similar to those of the other electrolytic depositions [35], having the following stages:

- electrochemical degreasing;
- washing in hot water and cold running water;
- pickling in sulfuric acid solution;
- washing in cold running water;
- pickling in potassium cyanide solutions (for silver or copper parts);
- washing in cold running water;
- gold plating;
- washing in cold distilled water (for electrolyte collection), then washing in cold running water;
- drying;
- technical control.

If the parts require a thorough decorative finishing before degreasing and after gold plating, the polishing operation is performed.

For gold plating, cyanuric electrolytes are used, which can be obtained either by anodic dissolution or by chemical dissolution, chlorinated electrolytes and ferrocyanide electrolytes [76,80].

Gold plating can be done on copper and copper alloy parts, as well as steel parts.

3.3.8. Cadmium plating

Cadmium is a white metal that has malleability and ductility properties similar to silver.

Cadmium depositions are used for technical corrosion protection, especially in contact with saltwater or chloride-based solutions, but due to high costs and toxicity [81], their use is reduced [4,5].

Sulfate, fluoroborate, cyanide and sulfamate electrolytes are used for cadmium coatings. The cadmium coating has good resistance to atmospheric corrosion, in marine climate, in contact with carbon dioxide, hydrogen sulfide, combustion gazelles and alkaline solutions. But it has low resistance to corrosion in distilled water, hydroxides, ammonia, chlorine, organic acids and minerals. To improve corrosion resistance, the cadmium coatings are passivated and dehydrogenated [82,83].

Other properties of cadmium coatings are low electrical contact resistance, superior ductility, ease of soldering, very good adhesion, malleability, high durability and lower hardness compared to zinc coatings etc.

Cadmium plating is used to coat steel, copper, aluminium and aluminium alloys. Cadmium plating is used to cover electrical parts and contacts mechanically stressed parts (springs, spark plugs, nuts, screws etc.) or to protect silver or nickel parts against blackening.

The technological process of cadmium deposition includes the following main stages: chemical degreasing, washing, chemical pickling, washing, cadmium plating, washing and technical control. If an improvement in the gloss of cadmium coatings is needed, after cadmium plating and washing, the parts are introduced into a 1% nitric acid solution or into chromic acid-based solutions. To increase the hardness, after cadmium coating, heat treatments can be applied to the parts, and to improve the corrosion resistance, it is recommended to apply thin layers of nickel and chromium [35].

3.3.9. Lead plating

Lead is a grey, heavy, malleable metal with a low melting point.

Because lead does not alloy with iron, it is necessary to use another metal that is easily alloyed with steel and forms a solid solution together with lead (e.g. tin). The coatings obtained are cheap, corrosion-resistant and with a smooth surface [84].

The lead coating has great advantages due to the high purity of the deposited metal. They are used to protect metals from the action of sulfuric acid, sulfurous water, sulfides and sulphates [85].

It is used in the chemical industry, in the military industry, in the automotive industry in the manufacture of gas tanks, exhaust drums, air filters. It is also used in fire extinguishers, boats etc.

3.3.10. Aluminium plating

Aluminium coatings are used in many fields, such as navigation, the aerospace industry, the electrical industry, the nuclear industry, the energy industry (manufacture of mirrors for capturing solar energy), etc.

The alloying is performed on iron alloys, copper and its alloys, as well as nickel and its alloys.

3.3.11. Iron plating

The iron coating has a silvery-white colour and has a compact microcrystalline structure. The properties of iron coatings are different from those of iron alloys, for example, it has a higher corrosion resistance compared to low carbon steel. The other properties differ depending on the electrolytes used, for example, if sulphate electrolytes are used, the structure of the iron coating will be more fragile compared to the iron coatings obtained with chloride-based electrolytes that have high plasticity [86].

Iron coatings are not used as a method of corrosion protection because they oxidize easily. Therefore, boiling is applied in:

- the printing industry to increase the wear resistance of lead stereotypes or printing plates;
- in the automotive and tractor industry for reconditioning used parts;
- for obtaining iron powders;
- coating hard alloy plates before gluing them to tools.

The technological process of iron plating to restore the dimensions of machine parts resembles that of other types of electrolytic deposits and includes the following stages: mechanical preparation (grinding or turning), insulation of parts not to be covered with bakelite varnish, perchlorvinyl lacquer etc.), mounting parts on the device, degreasing in alkaline solutions, washing, electrochemical pickling, washing, iron plating, washing, disassembly on the device, heat treatment of parts to dehydrogenate the part and change the mechanical characteristics of the coating and technical control [35].

3.3.12. Brass plating

Brass is a reddish-yellow copper-zinc alloy, depending on the percentage of copper and zinc. Brass coatings have higher hardness compared to copper coatings and high adhesion.

Brass plating is one of the oldest electrochemical deposition procedures. It is applied on steel, aluminium, zinc or as an intermediate layer for silver plating, nickel plating etc. [4,5].

In terms of corrosion resistance, brass coatings oxidize in air and do not resist the action of hydrogen sulfide. Brass is difficult to solubilize in hydrochloric acid and sulfuric acid and is easily solubilized in nitric acid. Cyanuric and pyrophosphoric electrolytes are used for brass plating [87].

The applications of brass coatings are:

- in the decorative industry - for the protection of various household objects, lighting fixtures, art objects etc.;
- in the machine-building industry for coating parts before rubber or those working in water vapour atmospheres;
- as an intermediate layer between steel and nickel, steel and silver, due to the high adhesion.

The technological process of brass plating contains the following steps: sanding, degreasing in lime or organic solvents, washing, brass plating, washing, drying, passivation or colouring of the coating [88].

3.4. Spray metal coatings

Spray metal coating is a thermomechanical process of coating a metal surface with another metal or alloy. This coating method can be used to deposit layers of aluminium, lead, nickel, tin, cadmium, copper and alloys (brass, coin, stainless steel, bronze, etc.).

In the case of this coating, the liquid metal is projected at high speed on the surface of the base material. The metal particles sprayed on the surface of the base metal form a layer due to the cohesive forces that appear between the metal molecules and due to the mechanical adhesion to the rough surface of the base material. The adhesion of the deposited layer depends very much on the spray speed of the metal and the surface condition of the base metal [89].

Depending on the material deposited on the base material surface, certain properties are improved, such as:

- Wear resistance;
- Resistance to high temperatures;
- Hardness;
- Corrosion and oxidation resistance;
- Thermal protection;
- Electrical insulation.

Due to its many properties, spray coating is used in many fields such as the electrical industry (capacitors, resistors), printing industry, medicine, jewelry and art industry, and for grinding parts.

Spray deposition has certain advantages and disadvantages that must be taken into account when choosing the deposition process. Among the advantages can be mentioned: high productivity, the possibility of covering certain areas of a part and of course the possibility of applying successive metal layers with different chemical compositions [89,90].

Regarding the disadvantages, in this method the material consumption is high, and the surfaces with deep holes or small concavity cannot be covered. Also, the deposited layer has an uneven thickness, with a porous structure, so there is the possibility of forming galvanic couplings between the deposited metal and the base metal, thus speeding up corrosion. Although productivity is high, the spraying process has a high cost, requiring equipment and installations for metallization and preparation of the base metal surface (by sandblasting) and also being required for the deposition of high purity materials. From the point of view of safety and health at work, as well as pollution, after spraying metal, metal dust is formed, high noise and toxic gases can be released, which involves taking measures for safety and health at work [4,5].

The deposition of the layers by spraying is done with the help of a spray gun, which aims to melt the deposited metal, spray and entrain the particles formed to the surface to be metallized, in which the metal is introduced in the form of wire or powder. To reach the liquid state, the metal is melted with the help of a flame, electric arc or plasma (allows the spraying of metal oxides and refractory materials with high melting point).

Spray coatings are made by different processes: thermal spray, arc spray, plasma spray, supersonic spray and detonation spray.

For the deposited layers to adhere very well to the surface of the base metal, it is necessary for the part to initially go through the surface preparation operations (degreasing, pickling and sandblasting). Given that the clean surface of the base material can oxidize quickly, it is important that immediately after the completion of the surface preparation, the spray deposition be carried out. Therefore, to avoid oxidation, there are automatic installations that perform electronically sandblasting and subsequently spraying the metal.

3.5. Diffusion coatings

Diffusion coating consists of covering the surface by changing the chemical composition of the base metal, to a certain depth, using an alloying element that diffuses into the surface layer.

The method consists in enriching the outer layer of the parts with atoms of corrosion-resistant metals or which lead to the formation, together with the atoms of the base metal, of structural constituents with properties superior to the base metal [91,92].

The method is based on the elementary processes that take place in the active environment of dissociation, adsorption and diffusion of atoms [93].

Aluminium, chromium, zinc and silicon layers can be deposited by this coating process (Table 3.1).

Table 3.1. Types of diffusion coatings.

Deposited metal	Process name	Properties	Deposition processes used	Base materials
Al	Aluminization	Oxidation resistance at high operating temperatures, 700-750 °C	in solid medium (powder mixtures) in molten Al baths.	low carbon steels mild carbon steels grey cast iron
Cr	Chromising	Superficial hardening Resistance to oxidation and corrosion at high temperatures Wear resistance	in solid medium in liquid medium in a gaseous medium vacuum chromising in fluidized layer	low carbon steels; alloys based on chromium, cobalt, nickel
Zn	Serardization	Resistance to atmospheric oxidation and corrosion	in zinc baths heated to a temperature of 400°C	steel cast iron
Si	Silicide	Oxidation resistance up to 870 °C Corrosion resistance in acidic media (HCl and H_2SO_4) Wear resistance High hardness	in silicon baths heated to a temperature of 700°C	low carbon steel cast iron

3.6. Cladding coatings

Cladding is a process of the permanent joining of two or more layered metals using cohesive forces and surface tension forces.

This is an effective method of protecting a metal with another corrosion-resistant metal. The cladding process combines the physical and mechanical properties of the base material with the anticorrosive properties of the deposited material (e.g. stainless steel deposited on carbon steel, gold-plated brass plating etc.). The adhesion of the protective layer is based on the diffusion under the simultaneous action of temperature and pressure [3].

The cladding process is done by casting, rolling, melting, welding or pressing on perfectly clean surfaces (degreased and pickled). Good adhesion between the two materials is obtained when they diffuse into each other. Compared to other types of coatings, the layers deposited by mechanical plating are thicker [4].

Metals that are used for cladding are aluminium, lead, nickel, copper, stainless steel etc.

The direct cladding process is done by casting the metal on the prepared surface (preheated) of the base material. The cladding piece is preheated. By solidifying the deposited metal on the surface of the base material, the adhesion between those two materials is achieved by diffusion to obtain a unitary laminated part. For example, copper-plated steel can be made by casting, the copper being heated to melting temperatures and then cast around the steel or the steel can be immersed in a molten copper bath. For example, copper-plated steel can be made by casting, the copper being heated to melting temperatures and then poured around the steel or the steel can be immersed in a molten copper bath. Thus, electrical conductors can be obtained, which have the properties of high mechanical strength of steel and the electrical conductivity of copper [5,94,95].

Press cladding is performed by pressing the cladding surfaces in such a way that the adhesion forces are necessary for a monolith piece to appear between them. Press cladding is done usually at high temperatures and under certain conditions at room temperature [96].

Roll cladding is done at temperatures corresponding to the ingot deformation. Aluminium cladding is done by rolling the steel, cleaning it and placing it between two aluminium sheets and cold or hot rolled for better adhesion of the aluminium layer deposited on the steel surface [97,98]. Also, hot rolling can be covered the carbon steel surface with a layer of stainless steel.

Weld cladding is applied to large bimetallic products (boilers, plates etc.). The metal can be deposited either by melting one or more wire or strip electrodes using the electric arc under a flux layer, on the surface of the base metal. This method is generally used to improve the abrasion resistance, for example for excavation tools [99,100].

References

[1]   Sundberg, P.; Karppinen, M. Organic and Inorganic–Organic Thin Film Structures by Molecular Layer Deposition: A Review. *Beilstein Journal of Nanotechnology 5:123* **2014**, *5*, 1104–1136. https://doi.org/10.3762/bjnano.5.123

[2]   Montemor, M.F. Functional and Smart Coatings for Corrosion Protection: A Review of Recent Advances. *Surface and Coatings Technology* **2014**, *258*, 17–37. https://doi.org/10.1016/j.surfcoat.2014.06.031

[3]   Florescu, A.; Bejinariu, C.; Comaneci, R.; Danila, R.; Calancia, O.; Moldoveanu, V. *Stiinta Si Tehnologia Materialelor*; Ed. Romanul: Bucuresti, 1997; Vol. II; ISBN 9739180469.

[4]   Udrescu, L. *Tratamente de Suprafata Si Acoperiri*; Politehnica: Timisoara, 2000; ISBN 9739389740.

[5]   Urdas, V. Tratamente Termice, Termochimice, Coroziunea Metalelor Si Acoperiri de Suprafata; Univ: Sibiu, 2001; ISBN 9736512991.

Advanced Coatings for the Corrosion Protection of Metals
Materials Research Foundations **115** (2022)

Materials Research Forum LLC
https://doi.org/10.21741/9781644901670

[6]   Faust, J.W. Studies on Surface Preparation. *Surface Science* **1969**, *13*, 60–71.
        https://doi.org/10.1016/0039-6028(69)90236-2

[7]   Ebnesajjad, S. Introduction to Surface Preparation and Adhesion. *Handbook of
        Adhesives and Surface Preparation* **2011**, 15–18. https://doi.org/10.1016/B978-1-
        4377-4461-3.10002-1

[8]   Pierre, C.; Bertrand, G.; Rey, C.; Benhamou, O.; Combes, C. Calcium Phosphate
        Coatings Elaborated by the Soaking Process on Titanium Dental Implants: Surface
        Preparation, Processing and Physical–Chemical Characterization. *Dental
        Materials* **2019**, *35*, e25–e35. https://doi.org/10.1016/j.dental.2018.10.005

[9]   Jamaati, R.; Toroghinejad, M.R. The Role of Surface Preparation Parameters on
        Cold Roll Bonding of Aluminum Strips. *Journal of Materials Engineering and
        Performance 2010 20:2* **2010**, *20*, 191–197. https://doi.org/10.1007/s11665-010-
        9664-7

[10]   Göhler, D.; Stintz, M.; Hillemann, L.; Vorbau, M. Characterization of
        Nanoparticle Release from Surface Coatings by the Simulation of a Sanding
        Process. *The Annals of Occupational Hygiene* **2010**, *54*.
        https://doi.org/10.1093/annhyg/meq053

[11]   Gong, H.; Pan, G.; Zhou, Y.; Shi, X.; Zou, C.; Zhang, S. Investigation on the
        Surface Characterization of Ga-Faced GaN after Chemical-Mechanical Polishing.
        Applied Surface Science **2015**, *338*, 85–91.
        https://doi.org/10.1016/j.apsusc.2015.02.107

[12]   Burduhos-Nergis, D.P.; Bejinariu, C.; Sandu, A.V. *Phosphate Coatings Suitable
        for Personal Protective Equipment*; Materials Research Forum LLC: Millersville,
        2021; Vol. 89; ISBN 9781644901113.

[13]   Senthil, K.; Mangalaraj, D.; Narayandass, S.K.; Adachi, S. Optical Constants of
        Vacuum-Evaporated Cadmium Sulphide Thin Films Measured by Spectroscopic
        Ellipsometry. *Materials Science and Engineering: B* **2000**, *78*, 53–58.
        https://doi.org/10.1016/S0921-5107(00)00510-9

[14]   Fedrizzi, L.; Rodriguez, F.J.; Rossi, S.; Deflorian, F.; di Maggio, R. The Use of
        Electrochemical Techniques to Study the Corrosion Behaviour of Organic
        Coatings on Steel Pretreated with Sol–Gel Zirconia Films. *Electrochimica Acta*
        2001, *46*, 3715–3724. https://doi.org/10.1016/S0013-4686(01)00653-3

[15]   Joshi, S.; Fahrenholtz, W.G.; O'Keefe, M.J. Alkaline Activation of Al 7075-T6 for
        Deposition of Cerium-Based Conversion Coatings. *Surface and Coatings*

Technology **2011**, *205*, 4312–4319. https://doi.org/10.1016/j.surfcoat.2011.03.073

[16]   Osakabe, S.; Adachi, S. Chemical Treatment Effect of (001) GaAs Surfaces in Alkaline Solutions. *Journal of The Electrochemical Society* **1997**, *144*, 290. https://doi.org/10.1149/1.1837397

[17]   Joshi, S.; Fahrenholtz, W.G.; OKeefe, M.J. Effect of Alkaline Cleaning and Activation on Aluminum Alloy 7075-T6. *Applied Surface Science* **2011**, *257*, 1859–1863. https://doi.org/10.1016/j.apsusc.2010.08.126

[18]   Rakowska, J.; Radwan, K.; Ślosorz, Z.; Porycka, B.; Norman, M. Selection of Surfactants on the Basis of Foam and Emulsion Properties to Obtain the Fire Fighting Foam and the Degreasing Agent. *Tenside Surfactants Detergents* **2014**, *51*, 215–219. https://doi.org/10.3139/113.110300

[19]   Luk, S.F.; Leung, T.P.; Miu, W.S.; Pashby, I. A Study of the Effect of Average Preset Voltage on Effective Case Depth During Electrolytic Surface-Hardening. *Materials Characterization* **1999**, *42*, 65–71. https://doi.org/10.1016/S1044-5803(98)00044-8

[20]   Luk, S.F.; Leung, T.P.; Miu, W.S.; Pashby, I. A Study of the Effect of Average Preset Voltage on Hardness during Electrolytic Surface-Hardening in Aqueous Solution. *Journal of Materials Processing Technology* **1999**, *91*, 245–249. https://doi.org/10.1016/S0924-0136(98)00441-5

[21]   Sivakumar, V.; Chandrasekaran, F.; Swaminathan, G.; Rao, P.G. Towards Cleaner Degreasing Method in Industries: Ultrasound-Assisted Aqueous Degreasing Process in Leather Making. *Journal of Cleaner Production* **2009**, *17*, 101–104. https://doi.org/10.1016/j.jclepro.2008.02.012

[22]   Besbes, S.; Ouada, H. ben; Davenas, J.; Ponsonnet, L.; Jaffrezic, N.; Alcouffe, P. Effect of Surface Treatment and Functionalization on the ITO Properties for OLEDs. *Materials Science and Engineering: C* **2006**, *26*, 505–510. https://doi.org/10.1016/j.msec.2005.10.078

[23]   Antony, O.A. Technical Aspects of Ultrasonic Cleaning. *Ultrasonics* **1963**, *1*, 194–198. https://doi.org/10.1016/0041-624X(63)90167-7

[24]   Bejinariu, C.; Burduhos-Nergis, D.-P.; Cimpoesu, N. Immersion Behavior of Carbon Steel, Phosphate Carbon Steel and Phosphate and Painted Carbon Steel in Saltwater. *Materials* **2021**, *14*, 188. https://doi.org/10.3390/ma14010188

[25]   Burduhos-Nergis, D.-P.; Vizureanu, P.; Sandu, A.V.; Bejinariu, C. Phosphate Surface Treatment for Improving the Corrosion Resistance of the C45 Carbon

Steel Used in Carabiners Manufacturing. *Materials* **2020**, *13*, 3410.
https://doi.org/10.3390/ma13153410

[26]   Burduhos-Nergis, D.P.; Vizureanu, P.; Sandu, A.V.; Bejinariu, C. Evaluation of
the Corrosion Resistance of Phosphate Coatings Deposited on the Surface of the
Carbon Steel Used for Carabiners Manufacturing. *Applied Sciences (Switzerland)*
2020, *10*. https://doi.org/10.3390/app10082753

[27]   Vynnycky, M.; Ipek, N. Electrochemical Pickling. *Lecture Notes in Computational
Science and Engineering* **2009**, *69 LNCSE*, 287–294. https://doi.org/10.1007/978-
3-642-00605-0_24

[28]   Kao, P.S.; Hocheng, H. Optimization of Electrochemical Polishing of Stainless
Steel by Grey Relational Analysis. *Journal of Materials Processing Technology*
2003, *140*, 255–259. https://doi.org/10.1016/S0924-0136(03)00747-7

[29]   Fotovvati, B.; Namdari, N.; Dehghanghadikolaei, A. On Coating Techniques for
Surface Protection: A Review. *Journal of Manufacturing and Materials
Processing 2019, Vol. 3, Page 28* **2019**, *3*, 28.
https://doi.org/10.3390/jmmp3010028

[30]   Alkire, R.C.; Kolb, D.M. Advances in Electrochemical Science and Engineering,
Volume 7. *Advances in Electrochemical Science and Engineering, Volume 7* **2001**,
7. https://doi.org/10.1002/3527600264

[31]   Lindsey, A.J. A Review of Electrolytic Methods of Microchemical Analysis.
Analyst **1948**, *73*, 67–74. https://doi.org/10.1039/an9487300067

[32]   Gerischer, Heinz. Mechanism of Electrolytic Deposition and Dissolution of
Metals. *Analytical Chemistry* **2002**, *31*, 33–39.
https://doi.org/10.1021/ac60145a007

[33]   Chen, R.; Trieu, V.; Schley, B.; Natter, H.; Kintrup, J.; Bulan, A.; Weber, R.;
Hempelmann, R. Anodic Electrocatalytic Coatings for Electrolytic Chlorine
Production: A Review. *Zeitschrift für Physikalische Chemie* **2013**, *227*, 651–666.
https://doi.org/10.1524/zpch.2013.0338

[34]   Abbott, A.P.; Barron, J.C.; Ryder, K.S. Electrolytic Deposition of Zn Coatings
from Ionic Liquids Based on Choline Chloride.
http://dx.doi.org/10.1179/174591909X438857 **2013**, *87*, 201–207.
https://doi.org/10.1179/174591909X438857

[35]   Corabieru, P.; Corabieru, A.; Vrabie, I. *Ingineria Suprafetelor. Depuneri Metalice
Prin Metode Electrochimice*; Tehnopress: Iasi, 2006; ISBN 9737023463.

[36] Eliaz, N.; Shmueli, S.; Shur, I.; Benayahu, D.; Aronov, D.; Rosenman, G. The Effect of Surface Treatment on the Surface Texture and Contact Angle of Electrochemically Deposited Hydroxyapatite Coating and on Its Interaction with Bone-Forming Cells. *Acta Biomaterialia* **2009**, *5*, 3178–3191. https://doi.org/10.1016/j.actbio.2009.04.005

[37] Aravinda, C.L.; Bera, P.; Jayaram, V.; Sharma, A.K.; Mayanna, S.M. Characterization of Electrochemically Deposited Cu–Ni Black Coatings. *Materials Research Bulletin* **2002**, *37*, 397–405. https://doi.org/10.1016/S0025-5408(01)00821-2

[38] Liu, W.; Qiao, X.; Liu, S.; Shi, S.; Liang, K.; Tang, L. A New Process for Pre-Treatment of Electroless Copper Plating on the Surface of Mica Powders with Ultrasonic and Nano-Nickel. *Journal of Alloys and Compounds* **2019**, *791*, 613–620. https://doi.org/10.1016/j.jallcom.2019.03.360

[39] Vladescu, A.; Vranceanu, D.M.; Kulesza, S.; Ivanov, A.N.; Bramowicz, M.; Fedonnikov, A.S.; Braic, M.; Norkin, I.A.; Koptyug, A.; Kurtukova, M.O.; et al. Influence of the Electrolyte's PH on the Properties of Electrochemically Deposited Hydroxyapatite Coating on Additively Manufactured Ti64 Alloy. *Scientific Reports 2017 7:1* **2017**, *7*, 1–20. https://doi.org/10.1038/s41598-017-16985-z

[40] Larson, C.; Smith, J.R. Recent Trends in Metal Alloy Electrolytic and Electroless Plating Research: A Review. **2013**, *89*, 333–341. https://doi.org/10.1179/174591911X13171174481239

[41] Xiang, T.; Ding, S.; Li, C.; Zheng, S.; Hu, W.; Wang, J.; Liu, P. Effect of Current Density on Wettability and Corrosion Resistance of Superhydrophobic Nickel Coating Deposited on Low Carbon Steel. *Materials & Design* **2017**, *114*, 65–72. https://doi.org/10.1016/j.matdes.2016.10.047

[42] Bautista, A.; González, J.A. Analysis of the Protective Efficiency of Galvanizing against Corrosion of Reinforcements Embedded in Chloride Contaminated Concrete. *Cement and Concrete Research* **1996**, *26*, 215–224. https://doi.org/10.1016/0008-8846(95)00215-4

[43] Kania, H.; Saternus, M.; Kudláček, J.; Svoboda, J. Microstructure Characterization and Corrosion Resistance of Zinc Coating Obtained in a Zn-AlNiBi Galvanizing Bath. *Coatings 2020, Vol. 10, Page 758* **2020**, *10*, 758. https://doi.org/10.3390/coatings10080758

[44] Gadzhov, I.; Lilova, D.; Ignatova, K. Membrane electrochemical preparation of three-polyphosphate non-cyanide electrolyte for coppering. *Journal of Chemical*

Technology and Metallurgy **2015**, *50*, 44–51.

[45] Jin, L.; Yang, J.-Q.; Yang, F.-Z.; Zhan, D.; Wu, D.-Y.; Tian, Z.-Q. Novel and Green Chemical Compound of HAu(Cys)2: Toward a Simple and Sustainable Electrolyte Recipe for Cyanide-Free Gold Electrodeposition. *ACS Sustainable Chemistry & Engineering* **2020**, *8*, 14274–14279. https://doi.org/10.1021/acssuschemeng.0c04438

[46] Sanches, L.S.; Domingues, S.H.; Marino, C.E.B.; Mascaro, L.H. Characterisation of Electrochemically Deposited Ni–Mo Alloy Coatings. *Electrochemistry Communications* **2004**, *6*, 543–548. https://doi.org/10.1016/j.elecom.2004.04.002

[47] Zečević, S.K.; Zotović, J.B.; Gojković, S.L.; Radmilović, V. Electrochemically Deposited Thin Films of Amorphous Fe–P Alloy: Part I. Chemical Composition and Phase Structure Characterization. *Journal of Electroanalytical Chemistry* **1998**, *448*, 245–252. https://doi.org/10.1016/S0022-0728(97)00417-8

[48] Tozar, A.; Karahan, I.H. Structural and Corrosion Protection Properties of Electrochemically Deposited Nano-Sized Zn–Ni Alloy Coatings. *Applied Surface Science* **2014**, *318*, 15–23. https://doi.org/10.1016/j.apsusc.2013.12.020

[49] Shacham-Diamand, Y.; Sverdlov, Y. Electrochemically Deposited Thin Film Alloys for ULSI and MEMS Applications. *Microelectronic Engineering* **2000**, *50*, 525–531. https://doi.org/10.1016/S0167-9317(99)00323-8

[50] Asgari, H.; Toroghinejad, M.R.; Golozar, M.A. On Texture, Corrosion Resistance and Morphology of Hot-Dip Galvanized Zinc Coatings. *Applied Surface Science* **2007**, *253*, 6769–6777. https://doi.org/10.1016/j.apsusc.2007.01.093

[51] Grandhi, S.; Raja, V.S.; Parida, S. Effect of Manganese Addition on the Appearance, Morphology, and Corrosion Resistance of Hot-Dip Galvanized Zinc Coating. *Surface and Coatings Technology* **2021**, *421*, 127377. https://doi.org/10.1016/j.surfcoat.2021.127377

[52] Peng, S.; Xie, S.K.; Xiao, F.; Lu, J.T. Corrosion Behavior of Spangle on a Batch Hot-Dip Galvanized Zn-0.05Al-0.2Sb Coating in 3.5 Wt.% NaCl Solution. *Corrosion Science* **2020**, *163*, 108237. https://doi.org/10.1016/j.corsci.2019.108237

[53] Al-Negheimish, A.; Hussain, R.R.; Alhozaimy, A.; Singh, D.D.N. Corrosion Performance of Hot-Dip Galvanized Zinc-Aluminum Coated Steel Rebars in Comparison to the Conventional Pure Zinc Coated Rebars in Concrete Environment. *Construction and Building Materials* **2021**, *274*, 121921.

https://doi.org/10.1016/j.conbuildmat.2020.121921

[54] Çetinkaya, B.W.; Junge, F.; Müller, G.; Haakmann, F.; Schierbaum, K.; Giza, M. Impact of Alkaline and Acid Treatment on the Surface Chemistry of a Hot-Dip Galvanized Zn–Al–Mg Coating. *Journal of Materials Research and Technology* **2020**, *9*, 16445–16458. https://doi.org/10.1016/j.jmrt.2020.11.070

[55] Jędrzejczyk, D.; Szatkowska, E. The Impact of Heat Treatment on the Behavior of a Hot-Dip Zinc Coating Applied to Steel During Dry Friction. *Materials 2021, Vol. 14, Page 660* **2021**, *14*, 660. https://doi.org/10.3390/ma14030660

[56] Jędrzejczyk, D.; Skotnicki, W. Comparison of the Tribological Properties of the Thermal Diffusion Zinc Coating to the Classic and Heat Treated Hot-Dip Zinc Coatings. *Materials 2021, Vol. 14, Page 1655* **2021**, *14*, 1655. https://doi.org/10.3390/ma14071655

[57] Sharma, A.; Ahn, B. Effect of Current Density and Alkaline PH on Morphology and Properties of Electroplated Sn. *Materials Research Express* **2020**, *6*, 126327. https://doi.org/10.1088/2053-1591/ab6543

[58] Sekar, R.; Eagammai, C.; Jayakrishnan, S. Effect of Additives on Electrodeposition of Tin and Its Structural and Corrosion Behaviour. *Journal of Applied Electrochemistry 2009 40:1* **2009**, *40*, 49–57. https://doi.org/10.1007/s10800-009-9963-6

[59] Deng, M.J.; Tsai, D.C.; Ho, W.H.; Li, C.F.; Shieu, F.S. Electrolytic Deposition of Sn-Coated Mesocarbon Microbeads as Anode Material for Lithium Ion Battery. *Applied Surface Science* **2013**, *285*, 180–184. https://doi.org/10.1016/j.apsusc.2013.08.032

[60] Li, C.F.; Ho, W.H.; Jiang, C.S.; Lai, C.C.; Wang, M.J.; Yen, S.K. Electrolytic Sn/Li2O Coatings for Thin-Film Lithium Ion Battery Anodes. *Journal of Power Sources* **2011**, *196*, 768–775. https://doi.org/10.1016/j.jpowsour.2010.07.068

[61] Santos, B.A.F.; Serenário, M.E.D.; Pinto, D.L.M.F.; Simões, T.A.; Malafaia, A.M.S.; Bueno, A.H.S. Evaluation of Micro-Crack Incidence and Their Influence on the Corrosion Resistance of Steel Coated with Different Chromium Thicknesses. *Revista Virtual de Quimica* **2019**, *11*, 264–274. https://doi.org/10.21577/1984-6835.20190019

[62] Li, J.; Li, Y.; Tian, X.; Zou, L.; Zhao, X.; Wang, S.; Wang, S.; Li, J.; Li, Y.; Tian, X.; et al. The Hardness and Corrosion Properties of Trivalent Chromium Hard Chromium. *Materials Sciences and Applications* **2017**, *8*, 1014–1026.

https://doi.org/10.4236/msa.2017.813074

[63] Lim, D.; Ku, B.; Seo, D.; Lim, C.; Oh, E.; Shim, S.E.; Baeck, S.H. Pulse-Reverse Electroplating of Chromium from Sargent Baths: Influence of Anodic Time on Physical and Electrochemical Properties of Electroplated Cr. *International Journal of Refractory Metals and Hard Materials* **2020**, *89*, 105213. https://doi.org/10.1016/j.ijrmhm.2020.105213

[64] Deepak, J.R.; Bupesh Raja, V.K.; Kaliaraj, G.S. Mechanical and Corrosion Behavior of Cu, Cr, Ni and Zn Electroplating on Corten A588 Steel for Scope for Betterment in Ambient Construction Applications. *Results in Physics* **2019**, *14*, 102437. https://doi.org/10.1016/j.rinp.2019.102437

[65] Viswanathan, R.; SIivashankar, N.; Chandrakumar, S.; Karthik, R. Improving Corrosion Resistance of Magnesium Alloy for Aerospace Applications . *International Journal of Mechanical and Production Engineering Research and Development* **2019**, *9*, 769–774. https://doi.org/10.1155/2020/4860256

[66] Tay, S.L.; Jadhav, P.; Goode, C. Hard Chrome Replacement with Cirrus Doped Electroless Nickel Coatings. *Key Engineering Materials* **2021**, *893*, 105–110. https://doi.org/10.4028/www.scientific.net/KEM.893.105

[67] Wang, Q.; Luo, S.; Wang, S.; Wang, H.; Ramachandran, C.S. Wear, Erosion and Corrosion Resistance of HVOF-Sprayed WC and Cr3C2 Based Coatings for Electrolytic Hard Chrome Replacement. *International Journal of Refractory Metals and Hard Materials* **2019**, *81*, 242–252. https://doi.org/10.1016/j.ijrmhm.2019.03.010

[68] Wang, Y.; Gu, Z.; Xin, Y.; Yuan, N.; Ding, J. Facile Formation of Super-Hydrophobic Nickel Coating on Magnesium Alloy with Improved Corrosion Resistance. *Colloids and Surfaces A: Physicochemical and Engineering Aspects* **2018**, *538*, 500–505. https://doi.org/10.1016/j.colsurfa.2017.11.055

[69] Laszczyńska, A.; Tylus, W.; Winiarski, J.; Szczygieł, I. Evolution of Corrosion Resistance and Passive Film Properties of Ni-Mo Alloy Coatings during Exposure to 0.5 M NaCl Solution. *Surface and Coatings Technology* **2017**, *317*, 26–37. https://doi.org/10.1016/j.surfcoat.2017.03.043

[70] Birlik, I.; Ak Azem, N.F. Influence of Bath Composition on the Structure and Properties of Nickel Coatings Produced by Electrodeposition Technique. *Journal of Science and Engineering* **2018**, *20*, 689–697. https://doi.org/10.21205/deufmd.2018205954

[71]   Jiang, N.; Liu, Y.; Yu, X.; Zhang, H.; Wang, M. Corrosion Resistance of Nickel-Phosphorus/Nano-ZnO Composite Multilayer Coating Electrodeposited on Carbon Steel in Acidic Chloride Environments. *Int. J. Electrochem. Sci* **2020**, *15*, 5520–5528. https://doi.org/10.20964/2020.06.50

[72]   Rahimi, E.; Rafsanjani-Abbasi, A.; Kiani-Rashid, A.; Jafari, H.; Davoodi, A. Morphology Modification of Electrodeposited Superhydrophobic Nickel Coating for Enhanced Corrosion Performance Studied by AFM, SEM-EDS and Electrochemical Measurements. *Colloids and Surfaces A: Physicochemical and Engineering Aspects* **2018**, *547*, 81–94. https://doi.org/10.1016/j.colsurfa.2018.03.045

[73]   Raghupathy, Y.; Kamboj, A.; Rekha, M.Y.; Narasimha Rao, N.P.; Srivastava, C. Copper-Graphene Oxide Composite Coatings for Corrosion Protection of Mild Steel in 3.5% NaCl. *Thin Solid Films* **2017**, *636*, 107–115. https://doi.org/10.1016/j.tsf.2017.05.042

[74]   Huang, J.; Tian, C.; Wang, J.; Liu, J.; Li, Y.; Liu, Y.; Chen, Z. Fabrication of Selective Electroless Copper Plating on PET Sheet: Effect of PET Surface Structure on Resolution and Adhesion of Copper Coating. *Applied Surface Science* **2018**, *458*, 734–742. https://doi.org/10.1016/j.apsusc.2018.07.119

[75]   Taghavi Pourian Azar, G.; Fox, D.; Fedutik, Y.; Krishnan, L.; Cobley, A.J. Functionalised Copper Nanoparticle Catalysts for Electroless Copper Plating on Textiles. *Surface and Coatings Technology* **2020**, *396*, 125971. https://doi.org/10.1016/j.surfcoat.2020.125971

[76]   Ashkenazi, D.; Inberg, A.; Shacham-Diamand, Y.; Stern, A. Gold, Silver, and Electrum Electroless Plating on Additively Manufactured Laser Powder-Bed Fusion AlSi10Mg Parts: A Review. *Coatings 2021, Vol. 11, Page 422* **2021**, *11*, 422. https://doi.org/10.3390/coatings11040422

[77]   Jiang, S.Q.; Newton, E.; Yuen, C.W.M.; Kan, C.W. Chemical Silver Plating and Its Application to Textile Fabric Design. *Journal of Applied Polymer Science* **2005**, *96*, 919–926. https://doi.org/10.1002/app.21541

[78]   Niece, S. la Silver Plating on Copper, Bronze and Brass. *The Antiquaries Journal* **1990**, *70*, 102–114. https://doi.org/10.1017/S0003581500070335

[79]   Okinaka, Y.; Hoshino, M. Some Recent Topics in Gold Plating for Electronics Applications. *Gold Bulletin 1998 31:1* **1998**, *31*, 3–13. https://doi.org/10.1007/BF03215469

[80]   Dimitrijević, S.; Rajčić-Vujasinović, M.; Trujić, V. Non-Cyanide Electrolytes for Gold Plating-A Review. *Int. J. Electrochem. Sci* **2013**, *8*, 6620–6646.

[81]   Flick, D.F.; Kraybill, H.F.; Dlmitroff, J.M. Toxic Effects of Cadmium: A Review. *Environmental Research* **1971**, *4*, 71–85. https://doi.org/10.1016/0013-9351(71)90036-3

[82]   Korpiniemi, H.; Huttunen-Saarivirta, E.; Kuokkala, V.T.; Paajanen, H. Corrosion of Cadmium Plating by Runway De-Icing Chemicals in Cyclic Tests: Effects of Chemical Concentration and Plating Quality. *Surface and Coatings Technology* **2014**, *248*, 91–103. https://doi.org/10.1016/j.surfcoat.2014.03.036

[83]   Baldwin, K.R.; Smith, C.J.E. Advances in Replacements for Cadmium Plating in Aerospace Applications. *http://dx.doi.org/10.1080/00202967.1996.11871127* **2017**, *74*, 202–209. https://doi.org/10.1080/00202967.1996.11871127

[84]   Silman, H.; Fry, M.F.E. The Lead Plating of Bronze Bearing Surfaces for High Pressure Fuel Pumps. **2017**, *23*, 43–58. https://doi.org/10.1080/00202967.1947.11869486

[85]   Iwanishi, H.; Hirose, A.; Imamura, T.; Tateyama, K.; Mori, I.; Kobayashi, K.F. Properties of Quad Flat Package Joints Using Sn-Zn-Bi Solder with Varying Lead-Plating Materials. *Journal of Electronic Materials 2003 32:12* **2003**, *32*, 1540–1546. https://doi.org/10.1007/s11664-003-0127-x

[86]   Thomas, J.D.; Klingenmaier, O.J.; Hardesty, D.W. Iron Plating—Process and Product Characteristics. **2017**, *47*, 209–216. https://doi.org/10.1080/00202967.1969.11870115

[87]   Clarke, S.G.; Bradshaw, W.N.; Longhurst, E.E. Studies on Brass Plating. **2017**, *19*, 63–89. https://doi.org/10.1080/00202967.1943.11869420

[88]   Saadatmand, M.; Sadeghpour, S.; Mohandesi, J.A. Optimisation of Brass Plating Condition in Plating of Patented Steel Wire. **2013**, *27*, 19–25. https://doi.org/10.1179/026708410X12459349719972

[89]   Hassani-Gangaraj, S.M.; Moridi, A.; Guagliano, M. Critical Review of Corrosion Protection by Cold Spray Coatings. **2015**, *31*, 803–815. https://doi.org/10.1179/1743294415Y.0000000018

[90]   Yin, S.; Chen, C.; Suo, X.; Lupoi, R. Cold-Sprayed Metal Coatings with Nanostructure. *Advances in Materials Science and Engineering* **2018**, *2018*. https://doi.org/10.1155/2018/2804576

[91]   Nicholls, J. Diffusion Coatings. *Anti-Corrosion Methods and Materials* **1966**, *13*,

21–23. https://doi.org/10.1108/eb006745

[92] Zhong, C.; Liu, F.; Wu, Y.; Le, J.; Liu, L.; He, M.; Zhu, J.; Hu, W. Protective Diffusion Coatings on Magnesium Alloys: A Review of Recent Developments. *Journal of Alloys and Compounds* **2012**, *520*, 11–21. https://doi.org/10.1016/j.jallcom.2011.12.124

[93] Castle, A.R.; Gabe, D.R. Chromium Diffusion Coatings. **2013**, *44*, 37–58. https://doi.org/10.1179/095066099101528216

[94] Jiang, Y.; Cheng, Y.; Zhang, X.; Yang, J.; Yang, X.; Cheng, Z. Simulation and Experimental Investigations on the Effect of Marangoni Convection on Thermal Field during Laser Cladding Process. *Optik* **2020**, *203*, 164044. https://doi.org/10.1016/j.ijleo.2019.164044

[95] Harada, H.; Nishida, S.I.; Suzuki, M.; Watari, H.; Haga, T. Direct Cladding from Molten Metals of Aluminum and Magnesium Alloys Using a Tandem Horizontal Twin Roll Caster. *Applied Mechanics and Materials* **2015**, *772*, 250–256. https://doi.org/10.4028/www.scientific.net/AMM.772.250

[96] Guo, X.; Fan, M.; Liu, Z.; Ma, F.; Wang, L.; Tao, J. Explosive Cladding and Hot Pressing of Ti/Al/Ti Laminates. *Rare Metal Materials and Engineering* **2017**, *46*, 1192–1196. https://doi.org/10.1016/S1875-5372(17)30135-2

[97] Naseri, M.; Reihanian, M.; Borhani, E. Bonding Behavior during Cold Roll-Cladding of Tri-Layered Al/Brass/Al Composite. *Journal of Manufacturing Processes* **2016**, *24*, 125–137. https://doi.org/10.1016/j.jmapro.2016.08.008

[98] Kim, I.K.; Hong, S.I. Mechanochemical Joining in Cold Roll-Cladding of Tri-Layered Cu/Al/Cu Composite and the Interface Cracking Behavior. *Materials & Design* **2014**, *57*, 625–631. https://doi.org/10.1016/j.matdes.2014.01.054

[99] Saha, M.K.; Das, S. Gas Metal Arc Weld Cladding and Its Anti-Corrosive Performance-A Brief Review. *Athens Journal of Technology and Engineering 5*, 155–174. https://doi.org/10.30958/ajte.5-2-4

[100] Venkateswara Rao, N.; Madhusudhan Reddy, G.; Nagarjuna, S. Weld Overlay Cladding of High Strength Low Alloy Steel with Austenitic Stainless Steel – Structure and Properties. *Materials & Design* **2011**, *32*, 2496–2506. https://doi.org/10.1016/j.matdes.2010.10.026

Chapter 4

Methods of Corrosion Protection by Depositing Inorganic and Organic Protective Layers

4.1. Introduction - inorganic protective layers

The process of coating with inorganic protective layers is based on the deposition of layers of inorganic compounds, chemically or electrochemically, by converting the surface metal layer into sparingly soluble metal products (oxides, phosphates, chromates, nitrides) or deposition of silicate-based masses or covering with ceramic materials. The quality of the coatings depends on the surface preparation and the working parameters [1,2].

Inorganic layers can be deposited by chemical or electrochemical conversion. The conversion layers result from the reaction between a reactant and the metal surface, forming a compounds film adhering to the surface, which ensures an increase in corrosion resistance and wears resistance, a low coefficient of friction, thermal and electrical insulation, relief of cold plastic deformation or can have a decorative role [3,4].

By chemical conversion, the metal surface reacts with an aqueous solution, resulting in a film of adherent and relatively compact corrosion products, which insulates the surface from the corrosive environment [5]. The chemical conversion processes are: phosphating, chromatization, sulfurization and misting.

The electrochemical conversion consists of the formation of a surface oxide layer of the base metal as a result of the reaction with oxygen resulting from electrolysis. The obtained layer has good corrosion-resistant, wear resistance and fatigue resistance [6]. Also, it has decorative properties and thermal and electrical insulation. The oxide formation processes are anodizing of aluminium, with Al_2O_3 formation; anodizing of other metals, with the formation of the oxide layer of the respective metal and electrolytic sulfurization with the formation of iron sulfide.

4.2. Phosphating

Phosphating is a process of coating metals, which consists in the formation on the metal surface a fine crystals layer of stable metal phosphates (iron, manganese or zinc) from aqueous solutions containing primary metal phosphates. These phosphates are hardly soluble in water, but soluble in mineral acids [7,8].

The coating can be presented under two physical aspects: the phosphatization itself, with a crystalline structure and the passivation, characterized by an amorphous coating.

Phosphate layers can be deposited on the surface of cast iron, steel, zinc, cadmium, magnesium, aluminium and their alloys.

The mechanism of the phosphate layer deposition process

For the formation of the phosphate layer, the most used metal ions are zinc, iron and manganese.

By introducing ferrous alloys (steel or cast iron) into the phosphating bath, following contact with free phosphoric acid or acid phosphates, the iron begins to dissolve according to the reaction (eq. 4.1):

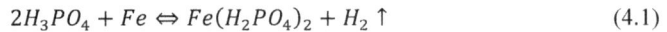

$$2H_3PO_4 + Fe \Leftrightarrow Fe(H_2PO_4)_2 + H_2 \uparrow \qquad (4.1)$$

The longer the reaction time between iron and phosphoric acid, the lower the concentration of acid, which leads to the enrichment of the solution with iron ions. At a certain concentration of phosphoric acid, the iron dissolution reaction takes place, and a thin film of iron phosphate is formed on the surface of the iron alloy, this layer having very low protection against corrosion [8,9].

To obtain a layer of phosphates that will provide the material with high protection against corrosion, high temperatures of the solution and a low acid concentration are required. Coating phosphates result from the reaction between metal and phosphoric acid, a reaction accompanied by the release of hydrogen, where the primary metal phosphates dissociate into secondary phosphates (eq. 4.2):[10]

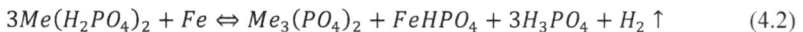

$$3Me(H_2PO_4)_2 + Fe \Leftrightarrow Me_3(PO_4)_2 + FeHPO_4 + 3H_3PO_4 + H_2 \uparrow \qquad (4.2)$$

where Me - ions of the metal used (zinc, iron, manganese etc.).

Phosphate layers properties

Depending on the metal ions used in the solution, as well as the chemical composition of the phosphating solution [10–12], the phosphate layer deposited on the metal surface may have the following properties:

- resistance to high temperatures (approx. 500 °C);
- high electrical resistance, being good electrical insulators;
- high adhesion capacity of oil and paint layers (used as a basis for future coatings);
- high porosity;
- corrosion resistance;
- crystalline or matte appearance;

- dark light grey or dark grey colour depending on the composition of the phosphating solution (dark grey - solution with high iron content, light grey - zinc phosphate layers).
- does not change the properties of the base material, such as elasticity, hardness, magnetic properties etc.;
- low frictional resistance, but by coating with lubricant the phosphate layer contributes to the decrease of coefficient of friction;
- low plasticity.

Applications of the phosphating process [8]

- protection against corrosion;
- decrease of the coefficient of friction due to the increase of the capacity of the materials to absorb lubricant: for covering some moving parts (bearings, axles etc.) or for plastic deformation (rolling, drawing, etc.);
- electrical insulation of stators, rotor plates, transformer plates etc.;
- for the preparation of the surface before painting, especially in the automotive industry;
- for local insulation of parts subjected to soldering, galvanizing or plumbing.

Phosphating processes

The processes for depositing phosphate layers are classified according to several criteria [8,10,13–16], which are presented in Fig. 4.1.

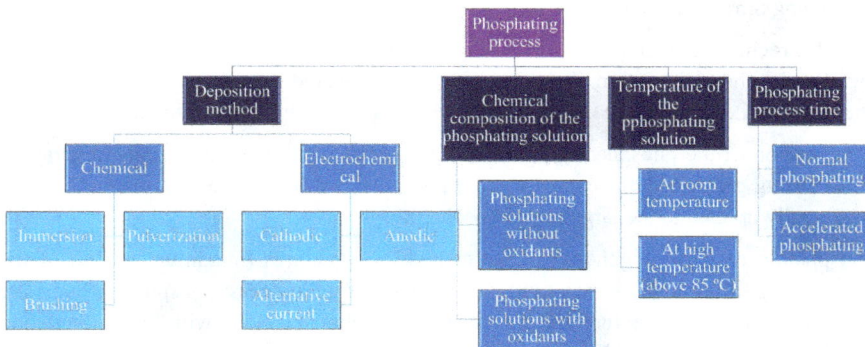

Figure 4.1. The classification of the phosphating process

Electrochemical phosphating is performed either with the help of direct current, the piece playing the role of a cathode, or passing through the bath an alternating current which, in the anodic stage dissolves iron and in the cathodic stage facilitates the release of hydrogen. The electrochemical process results in a deposition of a dense monocrystalline layer.

From the chemical composition of the phosphating solution point of view, phosphating solutions that do not contain oxidants have many disadvantages, among which: lower corrosion resistance of the obtained coating, short service life, consumes large amounts of chemicals and long process time.

By adding oxidants to the phosphating solutions, the above-mentioned disadvantages are eliminated. Among the oxidants used are chlorates, nitrates and nitrites that help to transform soluble primary ferrous phosphate into insoluble ferric phosphate, which is deposited on the bottom of the phosphate bath. Thus, by reducing the iron phosphate in the solution the corrosion resistance is improved. Also, due to the oxidants added to the phosphating solution, hard-to-attack metals can be covered and the phosphating process time is short (accelerated phosphating).

Regarding the phosphating process time, there are many methods to accelerate the deposition of the phosphate layer, such as chemical, electrochemical and mechanical methods.

The acceleration of phosphating by the chemical method is done by introducing in the phosphating solution some accelerating substances, such as oxidizing substances, reducing substances etc. These substances reduce the phosphating time by a few minutes, the most used being nitrates and chlorates.

The electrochemical method of accelerating the phosphating process consists in placing the metal under the action of an alternating current. Compared to the layers obtained by the chemical method, they have high porosity, being used mainly to cover small parts.

Acceleration of the phosphating process by the mechanical method consists in spraying under strong pressure the phosphating solution on the surface of the base material, which is generally applied to the large parts phosphating.

Phosphating of parts is generally done at high temperatures (maximum 98 °C) of the phosphating solution. However, to reduce the cost of the process, phosphating baths are also used at room temperature, the phosphate layers obtained being with finer granulation and lighter in colour, but with much lower corrosion resistance.

Stages of the phosphating process

The main steps of the phosphating process are shown in Fig. 4.2.

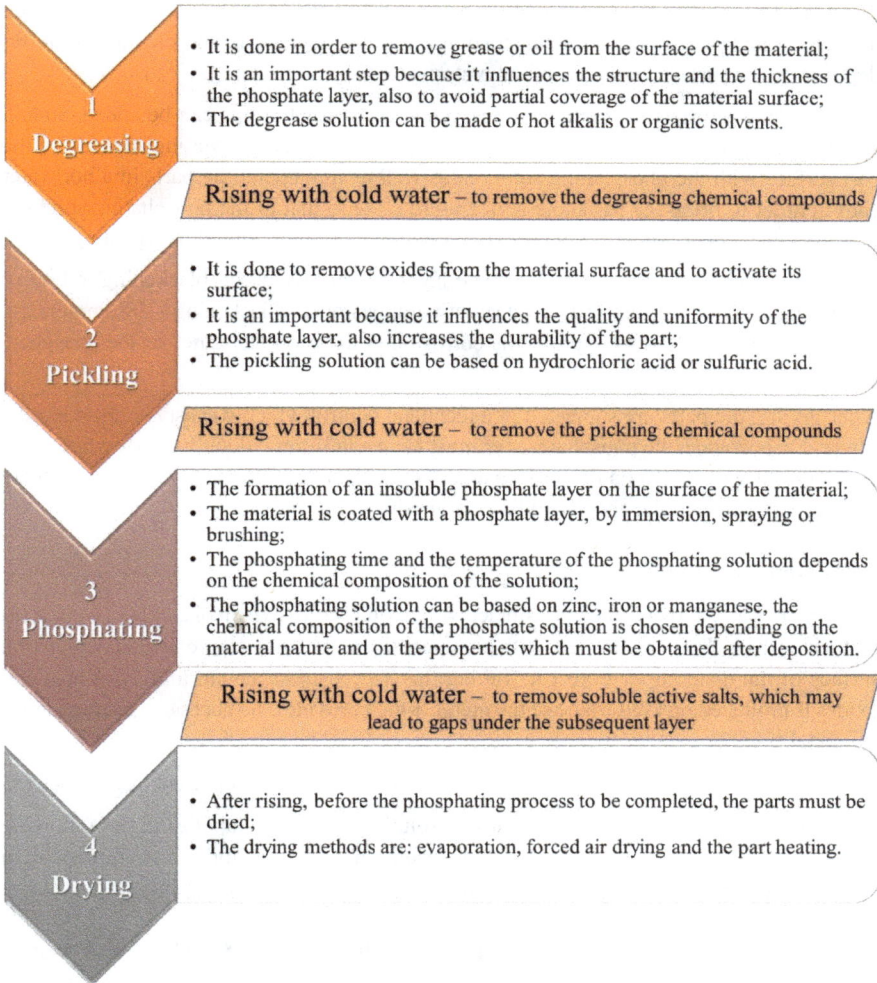

1 Degreasing
- It is done in order to remove grease or oil from the surface of the material;
- It is an important step because it influences the structure and the thickness of the phosphate layer, also to avoid partial coverage of the material surface;
- The degrease solution can be made of hot alkalis or organic solvents.

Rising with cold water – to remove the degreasing chemical compounds

2 Pickling
- It is done to remove oxides from the material surface and to activate its surface;
- It is an important because it influences the quality and uniformity of the phosphate layer, also increases the durability of the part;
- The pickling solution can be based on hydrochloric acid or sulfuric acid.

Rising with cold water – to remove the pickling chemical compounds

3 Phosphating
- The formation of an insoluble phosphate layer on the surface of the material;
- The material is coated with a phosphate layer, by immersion, spraying or brushing;
- The phosphating time and the temperature of the phosphating solution depends on the chemical composition of the solution;
- The phosphating solution can be based on zinc, iron or manganese, the chemical composition of the phosphate solution is chosen depending on the material nature and on the properties which must be obtained after deposition.

Rising with cold water – to remove soluble active salts, which may lead to gaps under the subsequent layer

4 Drying
- After rising, before the phosphating process to be completed, the parts must be dried;
- The drying methods are: evaporation, forced air drying and the part heating.

Figure 4.2.   The steps of the phosphating process.

Regardless of the method used for deposition of the phosphate coating, the surface should be previously prepared material. The quality and uniformity of the phosphate layer depend on how the surface was cleaned before phosphating [7,10].

The two steps used to prepare the surface for the deposition of the phosphate layer are degreasing and pickling. The chemical composition of degreasing and pickling solutions depends on the surface condition of the material [7,17].

After these two surface preparation steps, other intermediate steps can be added, among which: treating the parts in a soap and lye solution to increase the wetting capacity of the part surface with the phosphating solution, as well as immersing the parts in a hot water bath in order to bring the parts to a temperature close to that of the phosphating solution, thus avoiding a decrease in the phosphating rate (generally for normal phosphating).

Also, after drying the parts, the phosphate layer can be treated by blackening, oiling or varnishing in order to improve the decorative appearance (blackening - by immersion, spraying or brushing) or to improve the corrosion or abrasion resistance of the phosphate layer.

After drying, the parts are subjected to a technical control which consists in checking the external appearance, the phosphate layer must have a uniform colour, have no white spots (salts) or red spots (rust) on the surface and of course, on the surface of the piece so there are no areas in which the phosphate layer is missing.

4.3. Oxidation

When metals and alloys come into contact with various corrosive media, oxide layers form on their surface. They are uneven, with high porosity and do not adhere well enough to the surface of the material to protect it. Taking this aspect into account, a layer of artificial oxides is produced on the surface of the metal, which provides the necessary protection for the metal [9].

Artificial oxide layers can be obtained by several methods [9,18], in different media, such as:

- formation of oxide layers in aqueous solutions: generally applies to the following metals: magnesium, aluminium, zinc, cadmium and iron), formed by immersion in an aqueous solution whose chemical composition differs depending on the nature of the metal subject to oxidation;
- formation of oxide layers in the gas phase: the parts are heated in air or another gaseous environment, to increase the thickness of the layer already formed by oxides;

- formation of oxide layers by immersion in molten salts and metals: this process is applied to ferrous metals consisting in immersing the part in the melt for a few minutes at high temperatures, thus obtaining a black layer.

To protect the metal, the artificial oxide layers must have the following properties: good adhesion; uniform structure; low porosity, uniform thickness and good wetting capacity for oils and varnishes.

The main uses of oxidation are the protection of the metal against corrosion and their decoration (some oxide films may have different colours, for example, aluminium oxide layers are used in architecture and machine building or oxide layers obtained on steel are used in industry military, to weapons) [19].

4.3.1. Oxidation of ferrous metals (Blackening/Bluing)

Blackening/Bluing is the obtaining of iron oxides layers that aim to protect the ferrous metal against corrosion, especially if the part is subsequently oiled. Oxidation of ferrous alloys can be done by several methods:

a) Chemical blackening with alkalis

By this method, the oxide layer is formed by immersing the ferrous metal in concentrated alkaline solutions (alkaline hydroxide + oxidant), at high temperatures [20].

The steps of this method are degreasing, pickling or sandblasting, browning, drying, hot oil lubrication and technical control. After each step, the parts are washed to remove any residue on the metal surface.

If the oxide layer must be thicker, the blackening step can be performed by successively immersing the metal in two baths. Also, for the oxide layer formed on the surface to provide adequate protection against corrosion, in addition to browning, the phosphating of the part can be done [9].

b) Chemical blackening without alkalis

The oxide layer is formed by immersing the part in solutions based on metal chlorides (ferric chloride, mercury chloride etc.). This time, the surface of the ferrous metal is greased with chloride-based solutions and left to act for a few hours. Subsequently, in order to obtain the necessary colour, the parts are brushed, and in order to obtain a corresponding corrosion resistance, the parts are greased with oil [9].

The steps of this method are preparation of the surface by mechanical operations (sandblasting, sanding, polishing), lubrication of parts with metal chloride solution, drying, brushing, again lubrication with metal chloride solution, brushing and hot oil treatment [21].

c)   Electrochemical blackening

This method of forming the oxide layer on the surface of the metal using an electric current is used to obtain a high corrosion resistance. The protection layer is formed by immersing the metal in an alkaline solution, thus forming iron hydroxide which leads to the passivation of iron, slowing down the corrosion process. The anode is the metal immersed in the solution and the cathode is composed of coal, graphite and steel. Electrochemical browning is done in a short time and at a lower temperature compared to the others, but the biggest disadvantage of this method is the impossibility of browning parts with complicated configurations [9,22].

The oxide layer formed on the surface of the ferrous metal differs depending on the temperature at which the browning takes place and the chemical composition of the steel. For example, if the iron oxide is formed then the layer will have a red-brown colour, and if the ferrosoferric oxide is formed, then the colour will be black-bluish. Also, if the temperature at which the browning is performed will be low, the colour of the layer will be red, while at high temperature the colour of the layer will be blue. At the same time, for high silicon steels, the colour of the layer will be yellow-brown, and for alloy steels, the colour of the layer will be purple-red. In the case of cast irons, the colour of the layer will be golden-yellow to brown [22,23].

Browning is mainly used in the military industry in the manufacture of weapons, in the machine-building industry etc.

4.3.2. Chromating

Chromating is the process of forming protective layers of chromates on the surface of non-ferrous materials (aluminium, cadmium, zinc, copper etc.) and their alloys [24].

The process consists in immersing the parts in baths based on chromic or bichromatic acid, in the presence of an acid (sulfuric acid, phosphoric acid etc.). The chromium layer is obtained by dissolving the metal which occurs simultaneously with the reduction of hexavalent chromium ions. The obtained layer consists of alkaline chromates, has no porosity, is glossy and has good properties of corrosion resistance in atmospheric conditions [25].

Depending on the base metal, the colour of the obtained layer differs, for example in the case of zinc chromating the surface will have a silver colour with shades of blue, and in the case of aluminium chromating the layer will have a silver colour [18,26].

The properties of the chromate layer are gelatinous structure, good corrosion resistance in the atmosphere and relatively good in aggressive environments, low hardness, good electrical weldability etc.

Due to the toxicity of the chromate layer, chromating is not used in the food industry but is used in the machine-building industry, in the agricultural industry (agricultural machinery), in the civil construction industry, in the military industry (weapons) etc. [27].

4.3.3. Oxidation of aluminium (anodizing)

In the atmosphere, a thin oxide layer is naturally formed on the surface of aluminium, which offers good protection against corrosion. However, in order to improve the corrosion resistance, the aluminium oxide layer is hardened by chemical or electrochemical methods [9].

In the case of layers obtained by the electrochemical method, their properties are superior compared to those obtained by the chemical method.

Anodizing is the process of forming and growing the aluminium oxide film, for protective-decorative purposes [28,29].

The properties of aluminium oxide layers differ depending on the method of production used or the nature of the base material. These are:

- high hardness (the highest hardness is the layers obtained on pure aluminium using the electrochemical method)
- high adhesion (the layer is intimately bonded to the base material);
- high-temperature resistance (400 °c);
- electrical insulation capacity;
- great reflection;
- heat emission capacity;
- the deposited layers can be porous, thus having a high absorption capacity of different varnishes, lubricants etc.;
- high corrosion resistance.

The solutions used to obtain the layers of aluminium oxide are based on sulfuric acid, chromic acid, oxalic acid, boric acid etc. Electrochemical oxidation of aluminium and its alloys takes place by immersing the part in the anode bath, using direct or alternating current. Depending on the electrolyte used, the cathode can be made of lead (if the solution is based on sulfuric acid) and stainless steel or graphite (if the solution is based on chromic acid or oxalic acid). When the electric current passes through the solution, oxygen is released on the surface of the base metal, which reacts with the metal, forming a film of aluminium oxide [30,31].

Simultaneously with the formation of the oxide, its partial dissolution in solution takes place. Following the dissolution reaction, pores appear in the oxide layer, thus being possible to continue the reaction between metal and oxygen, leading to the further formation of the oxide layer, thus obtaining layers with large thicknesses.

Depending on the solution used and the chemical composition of the base metal, the properties and colour of the layer differ. For example, the layers of aluminium oxide obtained in sulfuric acid are colourless, and those that are obtained on alloys containing manganese, silicon, magnesium the layer will be grey or brown.

Oxidation in oxalic acid is performed for protective purposes and to improve mechanical properties (especially wear resistance).

Films obtained in an electrolyte based on chromic acid are thin, opaque, with low abrasion resistance, low porosity and good corrosion resistance. It is used to protect parts with a more complex configuration, as any electrolyte retention in the pores does not cause the corrosion of the base metal.

Anodizing is used in many areas [9,32], such as:

- in the food and consumer goods industry, for example, kitchen utensils, general-purpose parts;
- in architectures: architectural pieces or ornament pieces;
- in the construction of machines: mechanically stressed parts;
- in the manufacture of cellulose;
- in the textile industry;
- in the naval industry: naval equipment and components;
- in the aerospace industry;
- in the electrical and radio industry: material for reflectors of infrared heating installations;
- in the optical industry.

Although the chemical method is simpler and low cost, the electrochemical method is more often used due to the superior properties of the aluminium oxide layers obtained [33].

4.4. Enamelling

Enamel is a non-metallic material with high resistance to aggressive environment, good adhesion, thermal stability and resistance to mechanical stress [2].

Enamelling the deposition of a layer of molten glass on the surface of the metallic material, which can be applied on cast iron, steel, non-ferrous metals and noble metals.

In this case, a glass mass is used by melting natural rocks (sand, clay) with fluxes (borax, fluoride) and metal oxides (copper, nickel, cobalt, chromium) which gives the ceramic material colour and adhesion to the base metal. The adhesion of enamel to the base material is given by the oxide content of its composition, usually using cobalt oxide [34].

The chemical composition of enamel differs depending on the nature and properties of the base material (coefficient of expansion). To obtain a coefficient of expansion approximately with that of the base material, quartz and feldspar are melted, being mixed with fluxes such as borax, cryolite, calcined soda, fluorine, lead oxide etc. [19].

The enamel layer is generally deposited by immersing it in a water bath in which the finely ground ingredients corresponding to the chemical composition of the enamel used are introduced [35]. Of course, as with other types of coatings, the surface of the metal must be prepared by degreasing, pickling, and subsequently, if necessary, the parts are immersed in a nickel-based solution to improve the adhesion and thermal behaviour of the deposited layer. Then the pieces are washed and dried. After applying the enamel layer, the parts are dried, then being heated for $1 \div 4$ minutes at a temperature of about 815-870 °C and cooled to ambient temperature [18,19].

Because after the first enamel coating, the sample has a dark colour due to the cobalt oxide content, another layer of enamel is applied which has the role of opening or changing the colour and providing additional protection. Therefore, the part is quenched with another solution that differs in terms of chemical composition depending on the properties to be obtained, then dried and heated for about 3 minutes and allowed to cool.

Enamel coatings are mainly used in the aeronautic industry and the energy industry due to their refractory properties and resistance to thermal shock, but also the decorative industry, architecture etc. [34,36].

4.5. Introduction –organic protective layers

The organic materials used are polymers, elastomers, bituminous masses and asphalt. Their application on the surface to be protected is done by films, linings, coatings and wrappings [2].

Coatings with organic materials are generally used to prevent corrosion of the material, which can be achieved by several methods, such as: by an inhibitory effect, by forming a protective barrier between the metal and the corrosive environment and by preventing the appearance of galvanic torques [37,38]. In addition, organic coatings can have a decorative or functional role, having several properties, such as the ability to reflect, different colours, non-slip, fireproof etc.

The main organic coatings are of the type of paints, varnishes and enamels. But this category also includes plastic powder coatings, elastomer foil coatings and bituminous coatings [19].

The organic coating is a complex process, which contains several stages including preparation of the base metal surface, preliminary treatment, application of the base coating, application of the intermediate coating and application of the final coating.

The type and chemical composition of the organic coating differs depending on the type of base material and its use. For example, for the deposition of organic layers on the surface of tin-plated sheets, used in the food industry, it must be taken into account that the organic coating used must be compatible with the surface conditions imposed [39,40].

4.6. Painting or varnishing

Varnishes and paints are solutions or suspensions which, when applied to any surface, form a thin film whose thickness is measured in microns. In order to obtain the desired properties, different types of paints are applied successively, thus constituting a painting system. Thus, a whole range of products can be used: primer, putty, paint, enamel and varnish [2,41].

Varnishes are solutions of cellulosic derivatives, natural or synthetic resins in volatile organic solvents.

Paints are dispersed systems of pigments and fillers in various substances that form films, usually drying oils (linseed or tung oil) are used.

Pigments are produced in a powdery state that does not dissolve in water, film-forming substance or solvents. They are added to the paint because they provide several properties, such as colour, corrosion resistance, abrasion resistance, fire resistance, opacity, increased durability, etc. To provide satisfactory corrosion resistance, lead minimum, zinc chromate and basic lead silico-chromate are generally used [19].

Drying oils are produced from the seeds of plants such as soybeans, flax, sunflower etc. Over time, attempts have been made to replace drying oil with various types of synthetic resins [18,19]. There are several types of resins used, such as:

- Alkaline resins, which are used in the manufacture of maintenance paints and the final coatings of parts, due to their properties of resistance, gloss and colour preservation;
- Phenolic resins, which are used to cover tin-plated sheets to ensure their corrosion resistance when immersed in water or an atmosphere with high humidity;

- Vinyl resins are used for protection against corrosion of parts that are found in environments with high humidity or in environments where various chemicals are present, being used in water-based paint;
- Epoxy resins offer great protection against corrosion and are used in paints covering tanks for storing various chemicals or in tin-plated tin containers.
- Acrylic resins are used in varnishes and heat-reactive coatings for cars, appliances, etc., due to their transparent colour and high strength properties.

Coating with paint or varnish layers

a) Preparation of surfaces for painting or varnishing

Since the adhesion between the coating and the metal is an important point in any painting system, the preparation of the sample surface plays an important role in the coating process. Taking into account the stages that the material goes through to obtain the finished part and the way of storing them, on its surfaces there are fats, foreign bodies, oxides, etc. Thus, surface cleaning can be done by mechanical, manual and mechanized processes, by degreasing and pickling [9,42,43].

The mechanical, manual and mechanized processes by which the cleaning is done are:

- cleaning with hand tools: it is used to remove weakly adhering oxides and impurities with the help of wire brushes, abrasive cloths, trowels etc.;
- cleaning with mechanized tools: electric or pneumatic machines are used to which wire brushes, cloths or abrasive discs are attached;
- vibration cleaning: there is a chipping and a hardening of the surface;
- blasting: use sand or cast iron sand to remove dirt, rust, impurities etc.;
- cleaning with organic solvents: it is used to remove grease;
- steam cleaning: specialized equipment is used or with manual spray aggregates, to obtain a stronger cleaning effect, alkaline chemicals are added in the superheated water;
- flame cleaning: it is used to remove thick oxide stains and old paints;
- alkaline degreasing or cleaning with alkaline substances: performs the chemical transformation of substances on the surface of the material (eg grease);
- acid pickling: used to remove oxides from the metal surface, using sulfuric acid or hydrochloric acid.

b) Painting or varnishing

Depending on the material, the configuration and the dimensions of the piece to be painted, it is chosen by which method the paint or varnish layer will be deposited [44]. These methods are:

- brush coating: it is generally made for decorative protective coating;
- spray coating: approximately 80% of the parts in the transport industry are painted using the spray gun;
- immersion coating: it is used for parts with a complex configuration, these being immersed in a bath based on primer or paint.
- coating by electrophoresis: it is done by unloading on metal support the ionized films that are in the colloidal suspension, in water. Thus, it is formed with a continuous layer whose thickness depends on the quality of the paint and the electrolytic conditions.

c) Drying the paint or varnish layers

Drying is the last operation of the coating process with paint or varnish, being necessary to be carried out as soon as possible after application [18]. It can be done in three ways:

- convection drying: it is the most used drying method, using hot air. The equipment used must allow the drying temperature to be adjusted according to the material.
- radiation drying: energy sources such as methane gas or electricity are used to heat radiant surfaces.
- combined heating drying: used in the machine-building industry with the help of radiation-convection heating dryers (eg double-radiant ovens heated by flue gases and forced air recirculation.

4.7. Bituminous coatings

These types of coatings use bituminous resins, such as coal tar, synthetic or natural asphalt. These resins are in the form of high viscosity solutions and are applied in thin layers, resulting in thick films due to the introduction of fillers or in the form of hot melts, which are called enamels. Bituminous resins behave well in subterranean conditions and in contract with water, however, their durability in conditions of exposure to sunlight is reduced [2].

References

[1]   Benmalek, M.; Dunlop, H.M. Inorganic Coatings on Polymers. *Surface and Coatings Technology* **1995**, *76–77*, 821–826. https://doi.org/10.1016/0257-8972(95)02601-0

[2]   Florescu, A.; Bejinariu, C.; Comaneci, R.; Danila, R.; Calancia, O.; Moldoveanu, V. *Stiinta Si Tehnologia Materialelor*; Ed. Romanul: Bucuresti, 1997; Vol. II; ISBN 9739180469.

[3] Malucelli, G. Hybrid Organic/Inorganic Coatings Through Dual-Cure Processes: State of the Art and Perspectives. *Coatings 2016, Vol. 6, Page 10* **2016**, *6*, 10. https://doi.org/10.3390/coatings6010010

[4] Sidky, P.S.; Hocking, M.G. Review of Inorganic Coatings and Coating Processes for Reducing Wear and Corrosion. **2013**, *34*, 171–183. https://doi.org/10.1179/000705999101500815

[5] Brown, G.M.; Shimizu, K.; Kobayashi, K.; Thompson, G.E.; Wood, G.C. The Development of Chemical Conversion Coatings on Aluminium. *Corrosion Science* **1993**, *35*. https://doi.org/10.1016/0010-938X(93)90156-B

[6] S. Spinner, N.; A. Vega, J.; E. Mustain, W. Recent Progress in the Electrochemical Conversion and Utilization of CO2. *Catalysis Science & Technology* **2011**, *2*, 19–28. https://doi.org/10.1039/C1CY00314C

[7] Bejinariu, C.; Burduhos-Nergis, D.-P.; Cimpoesu, N. Immersion Behavior of Carbon Steel, Phosphate Carbon Steel and Phosphate and Painted Carbon Steel in Saltwater. *Materials* **2021**, *14*, 188. https://doi.org/10.3390/ma14010188

[8] Burduhos-Nergis, D.P.; Bejinariu, C.; Sandu, A.V. *Phosphate Coatings Suitable for Personal Protective Equipment*; Materials Research Forum LLC: Millersville, 2021; Vol. 89; ISBN 9781644901113.

[9] Corabieru, P.; Corabieru, A.; Vrabie, I. *Ingineria Suprafetelor. Depuneri Metalice Prin Metode Electrochimice*; Tehnopress: Iasi, 2006; ISBN 9737023463.

[10] Darband, G.B.; Aliofkhazraei, M. Electrochemical Phosphate Conversion Coatings: A Review. *Surface Review and Letters* 2017, *24*. https://doi.org/10.1142/S0218625X17300039

[11] Burduhos-Nergis, D.P.; Cazac, A.M.; Corabieru, A.; Matcovschi, E.; Bejinariu, C. Characterization of Zinc and Manganese Phosphate Layers Deposited on the Carbon Steel Surface. In Proceedings of the IOP Conference Series: Materials Science and Engineering; Institute of Physics Publishing, July 17 2020; Vol. 877, p. 012012. https://doi.org/10.1088/1757-899X/877/1/012012

[12] Nergis, D.P.B.; Cimpoesu, N.; Vizureanu, P.; Baciu, C.; Bejinariu, C. Tribological Characterization of Phosphate Conversion Coating and Rubber Paint Coating Deposited on Carbon Steel Carabiners Surfaces. *Materials Today: Proceedings* **2019**, *19*, 969–978. https://doi.org/10.1016/j.matpr.2019.08.009

[13] Simescu, F.; Idrissi, H. Effect of Zinc Phosphate Chemical Conversion Coating on Corrosion Behaviour of Mild Steel in Alkaline Medium: Protection of Rebars in

Reinforced Concrete. *Science and Technology of Advanced Materials* **2008**, *9*. https://doi.org/10.1088/1468-6996/9/4/045009

[14]   Guenbour, A.; Benbachir, A.; Kacemi, A. Evaluation of the Corrosion Performance of Zinc-Phosphate-Painted Carbon Steel. *Surface and Coatings Technology* **1999**, *113*, 36–43. https://doi.org/10.1016/S0257-8972(98)00816-0

[15]   Dhouibi, L.; Triki, E.; Salta, M.; Rodrigues, P.; Raharinaivo, A. Studies on Corrosion Inhibition of Steel Reinforcement by Phosphate and Nitrite. *Materials and Structures* **2003**, *36*, 530–540. https://doi.org/10.1007/BF02480830

[16]   Fedosov, S.; Roumyantseva, V.; Konovalova, V. Phosphate Coatings as a Way to Protect Steel Reinforcement from Corrosion. *MATEC Web of Conferences* **2019**, *298*, 00126. https://doi.org/10.1051/matecconf/201929800126

[17]   Burduhos-Nergis, D.P.; Vizureanu, P.; Sandu, A.V.; Bejinariu, C. Evaluation of the Corrosion Resistance of Phosphate Coatings Deposited on the Surface of the Carbon Steel Used for Carabiners Manufacturing. *Applied Sciences (Switzerland)* **2020**, *10*. https://doi.org/10.3390/app10082753

[18]   Urdas, V. Tratamente Termice, Termochimice, Coroziunea Metalelor Si Acoperiri de Suprafata; Univ: Sibiu, 2001; ISBN 9736512991.

[19]   Udrescu, L. *Tratamente de Suprafata Si Acoperiri*; Politehnica: Timisoara, 2000; ISBN 9739389740.

[20]   Y E N U Tachalam, R.; Kanagaraj, ; D; Nara W A N, Y.L.; Subramanian, R. CHEMICAL BLACKENING OF STAINLESS STEEL. *Bulletin of Electrochemistry* **1986**, *2*, 383–388.

[21]   Tennant, W.C. Zinc Oxide-Photosensitized Photolysis of Lead Chloride. *Journal of Physical Chemistry* **2002**, *70*, 3523–3528. https://doi.org/10.1021/j100883a026

[22]   Eckl, M.; Zaubitzer, S.; Köntje, C.; Farkas, A.; Kibler, L.A.; Jacob, T. An Electrochemical Route for Hot Alkaline Blackening of Steel: A Nitrite Free Approach. *Surfaces 2019, Vol. 2, Pages 216-228* **2019**, *2*, 216–228. https://doi.org/10.3390/surfaces2020017

[23]   Gao, C.; Qu, N.; He, H.; Meng, L. Double-Pulsed Wire Electrochemical Micro-Machining of Type-304 Stainless Steel. *Journal of Materials Processing Technology* **2019**, *266*, 381–387. https://doi.org/10.1016/j.jmatprotec.2018.11.018

[24]   Schuman, T.P. Protective Coatings for Aluminum Alloys. *Handbook of Environmental Degradation of Materials: Second Edition* **2012**, 503–538. https://doi.org/10.1016/B978-1-4377-3455-3.00017-1

[25] Kapranos, P.; Brabazon, D.; Midson, S.P.; Naher, S.; Haga, T. Advanced Casting Methodologies: Inert Environment Vacuum Casting and Solidification, Die Casting, Compocasting, and Roll Casting. *Comprehensive Materials Processing* **2014**, *5*, 3–37. https://doi.org/10.1016/B978-0-08-096532-1.00503-3

[26] Chen, X.B.; Easton, M.A.; Birbilis, N.; Yang, H.Y.; Abbott, T.B. Corrosion-Resistant Coatings for Magnesium (Mg) Alloys. *Corrosion Prevention of Magnesium Alloys: A volume in Woodhead Publishing Series in Metals and Surface Engineering* **2013**, 282–312. https://doi.org/10.1533/9780857098962.2.282

[27] Makhlouf, A.S.H. Current and Advanced Coating Technologies for Industrial Applications. *Nanocoatings and Ultra-Thin Films* **2011**, 3–23. https://doi.org/10.1533/9780857094902.1.3

[28] Abrahami, S.T.; de Kok, J.M.M.; Terryn, H.; Mol, J.M.C. Towards Cr(VI)-Free Anodization of Aluminum Alloys for Aerospace Adhesive Bonding Applications: A Review. https://doi.org/10.1007/s11705-017-1641-3

[29] Tsangaraki-Kaplanoglou, I.; Theohari, S.; Dimogerontakis, T.; Wang, Y.M.; Kuo, H.H.; Kia, S. Effect of Alloy Types on the Anodizing Process of Aluminum. *Surface and Coatings Technology* **2006**, *200*, 2634–2641. https://doi.org/10.1016/j.surfcoat.2005.07.065

[30] Wood, G.C.; O'Sullivan, J.P. The Anodizing of Aluminium in Sulphate Solutions. *Electrochimica Acta* **1970**, *15*, 1865–1876. https://doi.org/10.1016/0013-4686(70)85024-1

[31] Grubbs, C.A. Anodizing of Aluminum. *Metal Finishing* **1999**, *97*, 476–493. https://doi.org/10.1016/S0026-0576(99)80049-X

[32] Habazaki, H.; Shimizu, K.; Skeldon, P.; Thompson, G.E.; Wood, G.C.; Zhou, X. Effects of Alloying Elements in Anodizing of Aluminium. **2017**, *75*, 18–23. https://doi.org/10.1080/00202967.1997.11871137

[33] Martínez-Viademonte, M.P.; Abrahami, S.T.; Hack, T.; Burchardt, M.; Terryn, H. A Review on Anodizing of Aerospace Aluminum Alloys for Corrosion Protection. *Coatings 2020, Vol. 10, Page 1106* **2020**, *10*, 1106. https://doi.org/10.3390/coatings10111106

[34] Cannillo, V.; Sola, A. Different Approaches to Produce Coatings with Bioactive Glasses: Enamelling vs Plasma Spraying. *Journal of the European Ceramic Society* **2010**, *30*, 2031–2039. https://doi.org/10.1016/j.jeurceramsoc.2010.04.021

[35]   O', K.P.; Kenneth, F.•; Stanton, T. Optimisation of the Enamelling of an Apatite-Mullite Glass-Ceramic Coating on Ti 6 Al 4 V. https://doi.org/10.1007/s10856-011-4392-6

[36]   Buckton, D. Enamelling on Gold. *Gold Bulletin 1982 15:3* **1982**, *15*, 101–109. https://doi.org/10.1007/BF03214613

[37]   Kosec, T.; Legat, A.; Miloev, I. The Comparison of Organic Protective Layers on Bronze and Copper. *Progress in Organic Coatings* **2010**, *69*, 199–206. https://doi.org/10.1016/j.porgcoat.2010.04.010

[38]   Upadhyay, V.; Battocchi, D. Localized Electrochemical Characterization of Organic Coatings: A Brief Review. *Progress in Organic Coatings* **2016**, *99*, 365–377. https://doi.org/10.1016/j.porgcoat.2016.06.012

[39]   van der Wel, G.K.; Adan, O.C.G. Moisture in Organic Coatings — a Review. *Progress in Organic Coatings* **1999**, *37*, 1–14. https://doi.org/10.1016/S0300-9440(99)00058-2

[40]   Møller, V.B.; Dam-Johansen, K.; Frankaer, S.M.; Kiil, S. Acid-Resistant Organic Coatings for the Chemical Industry: A Review. **1998**. https://doi.org/10.1007/s11998-016-9905-2

[41]   Robinson, A.F.; Dulieu-Barton, J.M.; Quinn, S.; Burguete, R.L. Paint Coating Characterization for Thermoelastic Stress Analysis of Metallic Materials. *Measurement Science and Technology* **2010**, *21*, 085502. https://doi.org/10.1088/0957-0233/21/8/085502

[42]   Moncillo, M.; Feliu, S.; Galvan, J.C.; Bastidas, J.M. The Effect of Water Soluble Contaminants at the Steel/Paint Interface on the Durability of the Paint Coating. *Journal of the oil and colour chemists association* **1988**, *71*, 11–17.

[43]   Mezghani, S.; Perrin, E.; Vrabie, V.; Bodnar, J.L.; Marthe, J.; Cauwe, B. Evaluation of Paint Coating Thickness Variations Based on Pulsed Infrared Thermography Laser Technique. *Infrared Physics & Technology* **2016**, *76*, 393–401. https://doi.org/10.1016/j.infrared.2016.03.018

[44]   Akbarinezhad, E.; Rezaei, F.; Neshati, J. Evaluation of a High Resistance Paint Coating with EIS Measurements: Effect of High AC Perturbations. *Progress in Organic Coatings* **2008**, *61*, 45–52. https://doi.org/10.1016/j.porgcoat.2007.09.004

Chapter 5

Advanced Methods for Deposition of Protective Thin Layers

5.1 Introduction

Thin layers are obtained by condensing the coating material and depositing it on the surface of the base material (of any nature) in the form of atomic or molecular particles. Thus, the coating material initially undergoes a phase transformation to the state of gas or vapour. When choosing the material for deposition, the density, hardness, modulus of elasticity, coefficient of thermal expansion, thermal conductivity, melting temperature and working temperature must be known [1]. It is also important to know the nature of the chemical bond (metallic, covalent or ionic), the tendency to interact with other materials, chemical stability etc.

The methods and procedures for deposition of thin layers have been continuously improved and diversified, expanding their area of application in new areas of interest. There are two types of methods used to obtain thin layers: chemical methods and physical methods [2]. The classification of the main processes for deposition of thin layers are presented in Fig. 5.1. Thus, among the most important deposition procedures of thin layers are:

- CVD (Chemical Vapor Deposition) process, which takes place at high and moderate temperatures or assisted by the plasma;
- PVD (Physical Vapor Deposition) process, which takes place by thermal evaporation, cathodic spraying or ion plating.

The choice of a certain method depends on the requirements for the properties of the thin layer, the maximum temperature that the substrate can withstand, the compatibility of the process with the processes applied to the substrate before and after deposition, production costs, efficiency and large scale manufacturing [3].

5.2 PVD method

The layers deposited by the PVD process are obtained by condensing gaseous fluxes of atoms or molecules on the metal surface. PVD is a process of depositing a thin layer, in the gas phase, in which the source of the material is physically transformed in a vacuum, from a substrate without any chemical reaction involved [4].

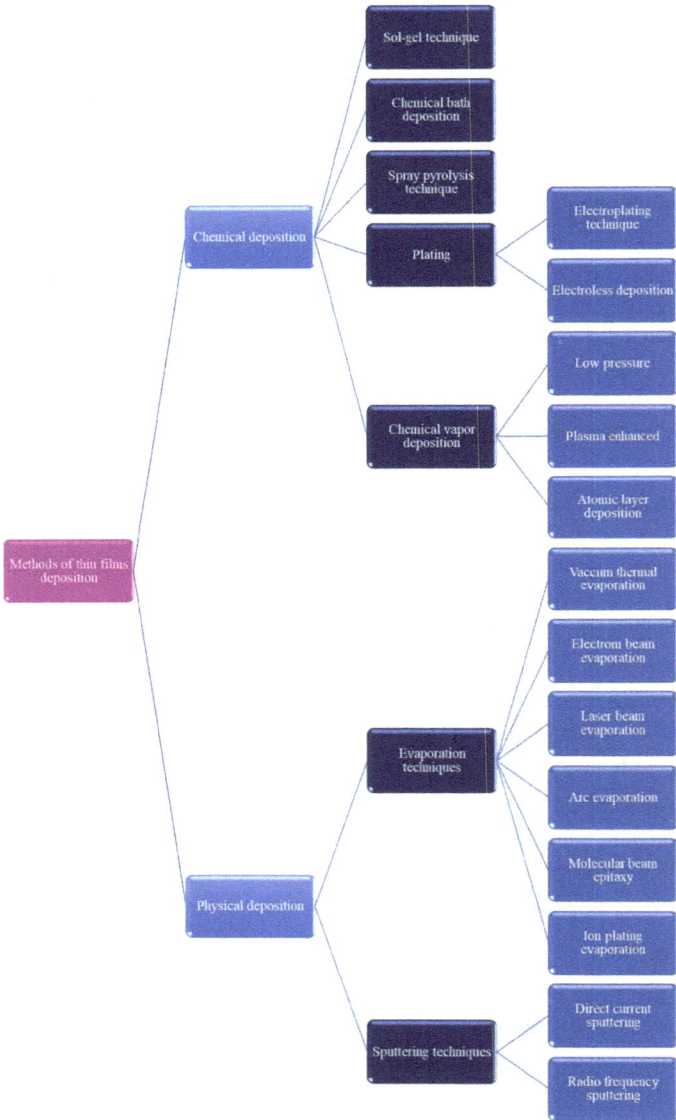

Figure 5.1.   Classification of the methods used for thin films deposition.

PVD deposition methods can be classified according to the formation of primary vapours in two categories: vaporization processes (thermal) and spray processes (kinetic) [3]. The layers obtained by this process have a thickness between $1 \div 10$ μm.

The deposition material may be in the solid or liquid state, so the passage of the vapour source atoms is done by physical mechanisms, i.e. by spraying, particle beam or evaporation.

Compared to other deposition methods, the PVD method has a multitude of advantages [5,6], such as:

- Allows precise control of the chemical composition and microstructure of the deposited layer;
- Layers of various materials can be deposited on any type of substrate that retains its properties at deposition temperatures;
- It is possible to obtain multilayer coatings with special properties of optical, magnetic, electronic, thermal etc.;
- The surface of the substrate is very little affected;
- Small loss of coating material;
- Higher corrosion resistance and high hardness compared to layers deposited by electroplating;
- Can be used for cutting very hard materials.

Although the method has many advantages, there are still some disadvantages [4,7] such as:

- Lack of ability to perform on-site cleaning on the substrate surface;
- The coverage step is difficult to correct;
- The need for high protection of staff, from the point of view of OSH;
- Cooling system required;
- High cost.

The main parameters to be considered when depositing through PVD are process temperature, pressure, particle energy and layer deposition speed.

Starting from the state of aggregation of the coating material and from the physical processes of generation, transport and nucleation of the deposition, the physical methods of deposition of thin layers in a vacuum can be classified into three basic types (thermal evaporation, spraying and ionic plating), from which derive a multitude of deposition methods [8]. When choosing a deposition method, eg PVD must take into account: the nature and properties of the coating material, the deposition rate, the adhesion of the layer to the base material, the properties that the deposited layer must-have, the thickness of the layer, the accuracy of deposition, the temperature of deposition, nature and properties of the base material, technological cost etc.

The applications of thin layers deposited in a vacuum can be systematized and presented synthetically based on the physico-chemical properties of the deposited films. Thus, through the three basic methods (thermal evaporation, cathodic spraying and ion plating), layers with various properties can be deposited on different materials (glass, plastics, ceramics or metals). The main applications of the layers deposited by the PVD method are presented in Table 5.1. [9–35].

Table 5.1. Applications of the layers deposited by the PVD method.

Layer type/ Properties	The main coating materials used	Applications
Conductive layers	Au, Ag, Pt, Ti, Cr, Cu, Cr/Cu/Au, Ti/Au, NiCr/Cu/Pd, Ti/Cu/Al	Contacts for integrated circuits
	Al or Zn deposited polypropylene film and paper capacitor SiO/Ta_2O_3, ZrO_2/Y_2O_3	Capacitors
	$Au-Ta_2O_5$, CrSiO, Au/Si	Cermet resistors
Resistive layers	NiCr, CrCo	High thermal stability resistors
	ITO, In_2O_3	Transparent heaters
	AlSi0.5...2 $AlCu_4Si$, TiN, W	Diffusion barriers
	ZrO_3; ZrO_2/N_2O_3; Ta_2O_5; SIO/Ta_2O_3	Capacitors
Decorative layers	Al - bright silver colour; Al + coloured varnishes - any colour; TiN_x, ZrN, HfN_x, TaC - yellow tones (gold to reddish and brown) TiC_x, HfC_x, ZrC_x, TaN - black tones (black-grey) CrN, Ta_2O_5, Mo_2C - white NB - dark blue, bright reddish SiO_2 - transparent - white.	Decorative coating with Al and coloured varnishes or SiO_2 for glass globes, ornaments, furniture, refrigerators etc. Decorative coating with TiN or Cu of flat glass, for making decorative mirrors Decorative coating with TiN (as a substitute for Au) for cases and bracelets of watches, cutlery, jewellery etc.
Magnetic layers	FeCr, Cr_2O_3 etc.	Magnetic discs - magnetic audio and videotapes
	NiFe, NiB, Gd-Co etc.	Magnetic heads
Semiconductor layers	InP - diodes, AlN and InSb - transistors in planar technology; $TiSi_2$ and WSi_2 - transistors in MOS technology	Optical components for electronics and optoelectronics
	GaP, ZnSMn, I_2O_3, SnO_2	LED
Superconducting layers	Nb etc.	Josephson junction

Optical layers	ordinary mirrors - Al, laser mirrors - Au, Ag, Al	Reflective films
Anticorrosive layers	Al, Cu, Cu-Zn, Al-Zn, 10TiNiCr180 etc.	Anticorrosive coating of steel parts and semi-finished products for ambient and normal temperature
	10 TiNi-Cr180, TiN etc.	Anticorrosive coating of steel sheet, for the food industry
	FeCrAlY, FeCrAl, CoCrAlY, NiCrAlTi	Anticorrosive and antioxidant coating, coating of turbine blades
Hard layers	Hard materials: TiB, TiC, TiN, ZrC, VC, TiAlN, TiNC etc.	Hardening of cutting tools (drills, taps, cutters, cutting plates, lathe knives etc.)
	Hard covalent material: B_4C, C, SiC –, SiB_6, Si_3N_4, AlN etc.	
	hard ionic (ceramic) materials: Al_2O_3, TiO_3, ZrO_2, MgO etc	
	TiN, TiC, TiAlN etc.	Hardening of pull-out mandrels and dies, cutting and stamping punches, raceways for bearings, high-wear machine parts (valves, cleats, camshafts, segments, pivots, pump bodies etc.)
Lubricating layers	Ag, Au, MoS_2, PTFE, TiN_x, TiC_x, TiN_xC_y, TiAlN, i-C, etc.	Hardening and reduction of the coefficient of friction for the running paths of the bearings, of the parts and of the machine parts that constitute friction couplings.

In addition to the areas presented in Table 5.1, the deposition of thin layers continuously increases its area of use, both in research works and in industrial applications, by replacing, in many cases, the inefficient and polluting classical methods of coating, or by creating of new applications. Thus, the thin layers of carbides and nitrides of transition metals characterized by high microhardness, with extremely low coefficient of friction and high melting point, can only be obtained by physico-chemical processes of vacuum deposition. The thin layers obtained by such vacuum deposition processes have extremely high adhesion to the substrate, comparable in size to the adhesion of a weld. The use of vacuum, as a medium for the deposition process, ensures that the deposited films, in addition to good adhesion, have a low impurity and an extremely low porosity and a reproducible composition, which gives the thin layers deposited in vacuum very good anti-corrosive properties. In addition, the methods of deposition of thin layers in a vacuum, offering the practical possibility of deposition of all metallic and non-metallic materials, in extremely varied combinations, thus obtaining anticorrosive films for all chemically aggressive environments, thus reducing the destruction and enormous losses of metals by corrosion.

Deposits of thin layers that use vacuum as a medium for the deposition process widen and continuously diversify their area of use. Among the new areas of use, anti-corrosion, lubricating and surface hardening deposits occupy a leading place.

5.2.1. Deposition of layers by thermal evaporation

Thermal evaporation deposition consists of vacuum heating of the coating material to a temperature higher than the melting temperature, followed by a condensation of the substrate vapours. By this method, thin layers are obtained, from simple substances, with a composition similar to that of the deposited layer. The particle energy of the vaporized material is small, about $0.1 \div 0.3$ eV [4].

As can be seen in Figs. 5.2, the deposition material is placed in the heating source and is transformed into vapours by heating it by Joule-Lenz effect. The basic principle of the process is the vaporization of the coating material and the formation of chemical reactions between the two materials.

Given that the area where the deposition takes place is vacuumed, there is a multitude of benefits, such as: lowering the vaporization temperature, removing impurities from the microcracks and pores of the substrate surface and facilitating the deposition on the substrate surface.

Although this method can be obtained layers with small thickness and the adhesion between the base material and the deposited layer is small, this method is the most used for functional and decorative coatings due to its simplicity and efficiency.

The quality of the deposit (its structure and porosity) depends very much on the flatness and roughness of the base material surface. Thus, if a thin layer is desired which has a metallic luster or is smooth, the surface of the base metal must be as rough as possible. Also, the temperature of the base metal greatly influences the structure of the layer (crystalline or amorphous) and the adhesion of the layer to the base material [37,38].

As in the case of other deposits, for the deposited layer to adhere as well as possible to the surface of the base metal, its surface must be prepared. Thus, it is envisaged to remove organic (fats) and inorganic (oxides) impurities from the surface of the base metal, as well as its activation.

Therefore, the preparation of the base metal surface is done in two stages. The first stage can be achieved by: degreasing or ultrasonic cleaning, reducing roughness and activating the surface by mechanical operations (sanding, sandblasting etc.). And the second stage is performed in the

Materials Research Forum LLC
https://doi.org/10.21741/9781644901670

deposition chamber by cleaning and activation by bombardment with the electron beam and plasma ion of the luminescent discharge, electron beam generated by the electron cannon; neutral ions or atoms generated by a generator of neutral ions or atoms [36,38].

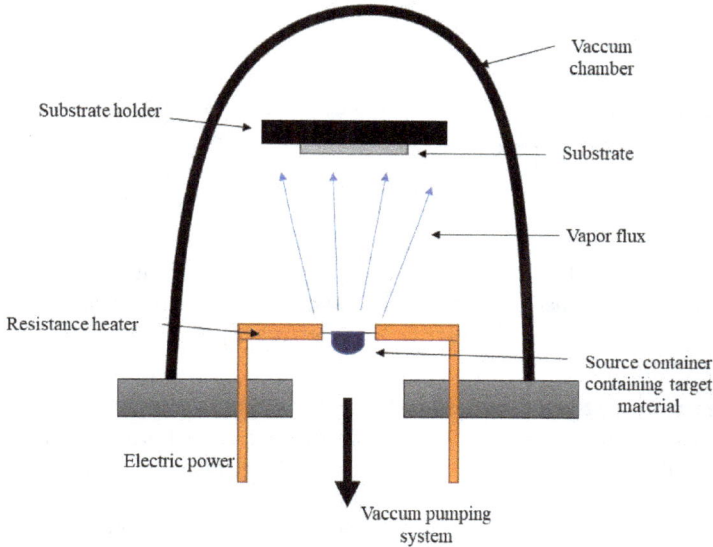

Figure 5.1.   The thermal evaporation deposition method [36].

Some of the advantages and disadvantages of the thermal evaporation deposition method [37–39] are listed in Table 5.2.

Table 5.2. Advantages and disadvantages of the thermal evaporation deposition method.

Advantages	Disadvantages
Versatile method	Layers with high porosity
High deposition rates at low temperatures of the base material	Maximum evaporation temperature ≈2000°C
Very high efficiency of use of the coating material	Possibility of decomposition of materials during evaporation
Stable mono/multilayer structures with uniform thickness can be obtained.	Inhomogeneities may occur in the chemical composition or layer structure (due to fluctuations in technological parameters)
Good adhesion of the layer to the base material	

5.2.2. Cathodic spray deposition

Spraying processes offer many advantages compared to other PVD or CVD processes, such as: improving the uniformity of thin layers, deposition of refractory materials, deposition of insulating films, very small targets, fine deposits, without microparticles as in the case of arc deposits [4,8].

Cathodic spraying consists in removing atoms from the surface of a plate from the deposition material fixed on the cathode of a continuous current source, as a result of its bombardment with positive argon ions accelerated in the electric field formed between the two electrodes (Fig. 5.2). The acceleration and increase of the energy of the positive ions are achieved, in this case, by polarizing the target to a negative potential of up to 5kV. The exploded atoms condense on the surface of the base material forming the coating [40,41].

For the deposition of chemical compounds (oxides, carbides etc.) it is introduced into the working chamber, together with the inert gas and reactive gas (nitrogen, metal gas, oxygen, etc.) which react with the target sprayed material to form on the surface of the substrate the respective compounds.

The parameters to be taken into account when cathodic spray deposition are: anode-cathode voltage, magnetic field induction, working pressure and current density in the discharge, respectively the average power dissipated in the discharge [42,43].

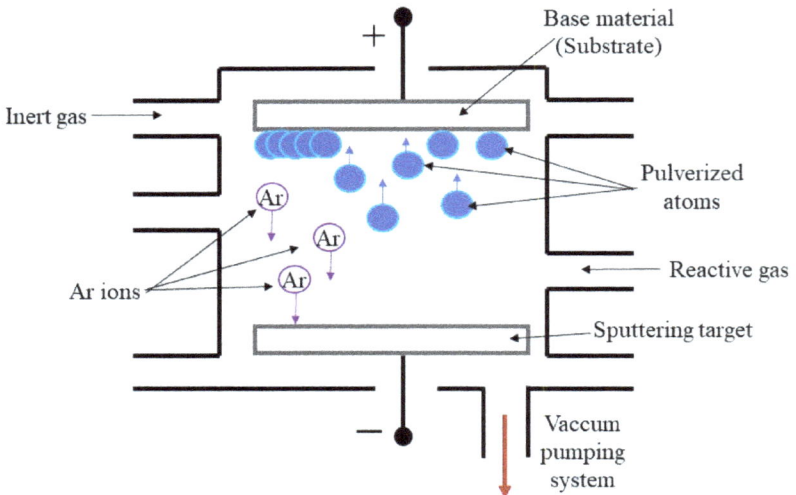

Figure 5.2.   Cathodic spray deposition [42].

The deposition rate of the layer differs depending on the working pressure. Thus, at pressures lower than 0.1 Pa and higher than 50 Pa the deposition speed is very low. It is also influenced by the density of the discharge current in direct proportion to its square and inversely proportional to the plate-substrate distance, which is about the centimetres.

Although the process of deposition of layers by spraying is simple and allows the deposition of several types of materials, it also has certain disadvantages related to the conditioning of the type of material (electroconductive materials - in the case of direct current spraying), the deposition rate (generally low) and low energy efficiency. To be able to be used for deposition and dielectric materials, the spray method in alternating current with radio frequency will be used.

Some of the advantages and disadvantages of the cathodic spray deposition method [2,4,41] are shown in Table 5.3.

Table 5.3. Advantages and disadvantages of the cathodic spray deposition method.

Advantages
The process used for the deposition of magnetic layers, the production of cryogenic films, the deposition of hard fusible materials
Can be deposited different types of alloys and compounds
The thickness uniformity of the deposited layer is very good, due to the plane-parallel geometry
Covering substrates with irregularities, especially if they are polarized with negative potential (from -10 V to -100 V)
High adhesion of the deposited layer, due to the cleaning of the substrates by ion bombardment
Disadvantages
Low deposition rate in conditions of a high voltage (over ~ 1000 V), respectively at high pressures (over ~ 1 Pa) of the discharge gas atmosphere
Expensive equipment/installations
Deposit material is available in the form of tiles.

5.2.3. Deposition of layers by ion plating

Among the three methods of deposition of thin layers by PVD, for obtaining coatings with special mechanical properties, ionic plating is mainly used, as it has a high degree of energetic activation of the substrate, which ensures a special adhesion to the films and a dense structure with very small grains [44].

The process consists in ionically bombarding the surface of the base metal both before deposition, for its cleaning by cathodic spray and its uniform heating to the deposition temperature, both phenomena having an effect on the increase of adhesion. Ionic bombardment during the deposition of the layer has the following effects [45]:

- Spraying weakly bonded atoms, gases and impurities;
- Favouring secondary nucleation;
- Increasing the mobility of condensed atoms;
- Activation of chemical reactions at the surface.

These effects produce important changes in the properties of the layers deposited by ion plating, compared to those obtained by thermal evaporation or spraying. Thus, the deposited layers have a finer structure, reduced porosity, high density and better adhesion to the substrate [46,47].

In general, for ion plating, in addition to the process with cathodic magnetron spraying, the electric arc evaporation deposition process is used for layer deposition (Fig. 5.3). It can be seen that the atoms of the deposition material, obtained thermally by heating, pass through the electric arc resulting from the discharge in argon, which takes place between the lateral electrodes. Thus, they ionize and condense on the surface of the base metal.

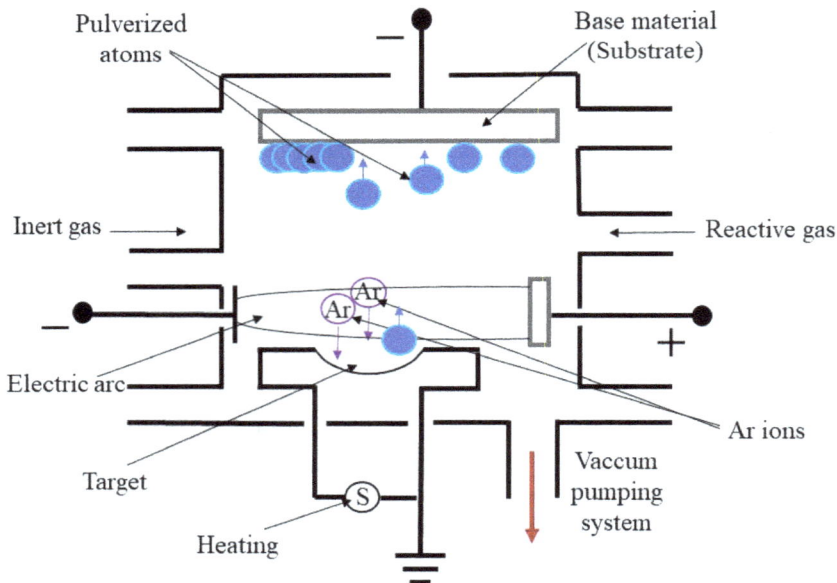

Figure 5.2. Arc ion plating deposition [48].

Ion bombardment has a complex influence on the properties of the deposited layers (structure, composition, compactness, etc.) by the average energy of the ions at the time of impact, by the ratio between the number of ions and neutral atoms reaching the surface of the substrate per unit time and by the angle under which the flow of ions bombards the substrate [44,46].

Some of the advantages and disadvantages of the ion plating method [44,45] are shown in Table 5.4.

Table 5.4. Advantages and disadvantages of ion plating method.

Advantages
High deposition speeds
The vaporized material is ionized in a proportion of 30-90%
Dense, uniform layers with good adhesion and high resistance to wear and corrosion are obtained
Very good quality of single / multilayer structures can be obtained at relatively low processing temperatures (<100°C)
Several metal compositions can evaporate without changing the composition of the remaining solid target
Cathodes can be placed in any position (horizontal, vertical, upside down), which makes possible a flexible design of the deposition installation
Disadvantages
A limited number of target materials can be used (oxide-free metals only, which do not have too low an evaporation temperature)
Due to the high current densities, a certain amount of the target material is ejected as small liquid drops.

5.3. CVD method

The chemical methods used for the deposition of thin layers of protection are based on different chemical processes, such as chemical reactions, atomic diffusion and chemical coatings. The CVD method is a chemical process that consists in passing over the surface of the base metal, heated in a sealed chamber (up to the vaporization temperature), a mixture of active gases that contains chemical elements of the layer in the form of volatile compounds. The layer deposited on the surface of the base material is formed as a result of complex chemical reactions that take place between the target, the gases in the mixture and the metal surface of the base material [4,8].

Almost any type of material can be deposited through CVD, but most of the time high temperatures and sometimes toxic gases are required, which are then difficult to remove. The conditions required for the coating are that its structure does not change irreversibly at the deposition temperature, has sufficient corrosion resistance to the gases used in the process and is compatible with the material of the coating [49].

In order for the layer to be deposited by the CVD method, the base metal must comply with the following conditions [49,50]:

- For the adhesion between the layer and the substrate to be as good as possible, the surface of the material must be smooth and clean;
- For the best possible corrosion resistance, the structure of the base material must not change at the temperature at which the deposition process takes place;
- It must be controllable with the deposited layer (eg approximately equal coefficients of expansion).

The main advantage of the CVD method is the possibility to cover parts with complex geometry. However, by this method, refractory ceramic layers and heavier metal layers of metals with low melting temperatures are deposited. However, the biggest disadvantage of this type of coating is the dangers caused by gases (toxic, explosive, corrosive etc.) [50,51].

Of all the CVD processes, the Plasma enhanced process has the most widespread use. In this process, the reaction activation energy is given by the thermal component and the plasma of the luminescent discharge in direct current or radiofrequency current.

The layers deposited by the CVD method can be:

- Made of metals and metal alloys;
- Of intermetallic compounds: carbides, oxides, phosphides, silicides, nitrides etc .;
- Made of extra hard materials: diamond type carbon, boron nitride etc.

Depending on the type of CVD process and the parameters used during the deposition process, different types of layers can be obtained, such as amorphous layers, layers with caps, layers with faceted structure, layers with columnar structure and monolithic layers [4].

Depending on the type of material deposited, the layers have the following properties: hardness (TiC, etc.), adhesive wear resistance (TiN, TiC, etc.), high durability (TiC / Al_2O_3), oxidation resistance (chromium carbide, etc.) etc.

The CVD layer deposition method is used in many fields, such as the medical industry, optical industry, aerospace industry, microelectronics, mechanical processing and plastic deformation processing, electrical industry, machine-building industry, etc. Some of the advantages and disadvantages of the CVD [49,50] method are shown in Table 5.5.

Table 5.5. Advantages and disadvantages of the CVD method.

Advantages
High wear resistance
It does not require vacuum or unusual levels of electricity
Selective deposition of layers
It is not needed to rotate the substrate (the part)
Suitable for parts with complex configurations
Disadvantages
The reactants or gases of the product are usually toxic, flammable, corrosive substances - ecologically and OSH problematic
High processing temperature
The thickness of the deposited layer is not uniform

5.4   Deposition of anticorrosive thin layers

The anti-corrosion protection of low-alloy steel by thin-film deposition in a vacuum, for many practical applications, has already passed from the stage of laboratory research to that of widespread industrial use. If the first industrial applications were based on one-component metal coatings (Al, Zn, Cd, Ti etc.), now coatings with multiple metal layers and especially with complex compounds, made of stainless steels or metals or non-metals (nitrides, carbides) are being expanded, silicones etc.). An eloquent example of the industrial application of vacuum thin-film deposits is the anticorrosive protection of steel sheets with one-component metal films (Al, Ti, Cu, Sn, Ni), metal alloy films (Ti, stainless steel etc.) by electron beam evaporation [11,12,17].

In addition to these industrial uses of vacuum-corroded anticorrosive films, the results of studies and research on new anticorrosive films and their applications are increasingly published in the technical literature.

Next, some of the results of the corrosion studies on some representative films, deposited in vacuum by magnetron spray, thermal evaporation or ion plating, will be summarized. High adhesion of deposits is obtained only at high substrate temperatures (generally above 300 °C ... 400 °C). To improve adhesion, the substrate is cleaned and activated by ion bombardment. Depending on the influence of luminescent discharge on adhesion to the metal substrate, the metals were divided into three groups [52–56].

The first group includes metals (Ca, Mg, Hf, Al, Zr, Ti, Si, Ta, Nb, Cr, V, In, Fe, Mo, Ge) for which the luminescent discharge brings significant increases in adhesion.

The second group includes mâetals (Mn, Cd, Ni, Co, Pr, Cu) for which their adhesion increases, but not significantly, due to the use of luminescent discharge.

The third group includes metals (Zn, Sn, Sb, Bi, Te, Pt, Au, Ag), for which the adhesion does not change due to luminescent discharge.

To improve the adhesion of Zn or Cd deposits on steel, the deposition of an intermediate layer of Sn or Pb can be used, which ensures good adhesion to the substrate.

5.5 Deposition of hard, lubricating and wear-resistant thin layers

The deposition of hard, lubricating and wear-resistant thin layers by physical processes is a modern technology with an extremely high economic efficiency [57–59]. It is used to:

- cutting tools coating (drills, cutters, lathe knives, cutting plates etc.);
- mandrels and moulds coating;
- car parts coating (cleats, valves, camshafts, segments, pivots etc.).

Today, a multitude of hard materials are known, but for a correct choice, several aspects must be taken into account. This is not easy, as the requirements for the substrate and the films are often very complex and sometimes antagonistic. For example, high adhesion to the substrate, lack of interaction of the film with the substrate, high hardness and strength of the film, are properties that can not be achieved simultaneously [16,31]. The criteria for selecting materials for wear-resistant coatings must take into account the following essential aspects:

- adhesion and interaction (reaction) of the substrate with the film take place at the film-substrate interface;
- the composition and microstructure of the film, influences the microhardness, the residual internal stresses, the resistance, the thermal stability of the film etc.;
- behaviour towards the environment or the part in contact is determined by the outer surface of the deposited film;

Hard, wear-resistant materials can be classified into three categories, depending on the nature of their chemical bond, namely: hard ceramic materials such as boron, carbide and metal nitride, covalent hard materials, including borons, carbides and nitrides of Al, Si and B, as well as diamond-type carbon and hard ceramic (ionic) materials, which include oxides of Al, Zr, Ti and Be [17,29,31,32].

Materials Research Forum LLC
https://doi.org/10.21741/9781644901670

5.6 Decorative thin film deposition

The deposition of thin layers in a vacuum with a decorative role, by physical and physico-chemical methods, are today widely used in the consumer goods industry with significant economic effects [15,16].

The first applications of thin layers deposited in vacuum to make decorative coatings, used the vacuum aluminization of plastic objects and parts and their subsequent coating with variously coloured or colourless varnishes, to ensure a more aesthetic appearance of the product [60]. This technology is used in the consumer goods industry, for making toys, jewellery, ornaments, ornaments, paper and metallic plastic foils, packaging etc.

Subsequently, the use of metal materials other than aluminium began to be used to make decorative coatings [61]. Due to a limited range of colours offered by simple metallic materials, the use of transition metal compounds (nitrides, carbides and oxides) has been used, which allow the production of variously coloured and highly ethical films [53,60]. For decorative coatings are used:

- titanium, zirconium or hafnium nitride, which provides colours from bright yellow to gold, reddish-brown, to brown to black, depending on the nitrogen content;
- titanium, zirconium or hafnium carbide, which provides colours from glossy grey to black depending on the carbon content;
- titanium oxide, zirconium or hafnium, which provides colours from azure to rose pink, depending on the oxygen content.

References

[1]   Florescu, A.; Bejinariu, C.; Comaneci, R.; Danila, R.; Calancia, O.; Moldoveanu, V. *Stiinta Si Tehnologia Materialelor*; Ed. Romanul: Bucuresti, 1997; Vol. II; ISBN 9739180469.

[2]   Corabieru, P.; Corabieru, A.; Vrabie, I. *Ingineria Suprafetelor. Depuneri Metalice Prin Metode Electrochimice*; Tehnopress: Iasi, 2006; ISBN 9737023463.

[3]   Kern, W.; Schuegraf, K.K. Deposition Technologies and Applications: Introduction and Overview. *Handbook of Thin Film Deposition Processes and Techniques* **2001**, 11–43. https://doi.org/10.1016/B978-081551442-8.50006-7

[4]   Udrescu, L. *Tratamente de Suprafata Si Acoperiri*; Politehnica: Timisoara, 2000; ISBN 9739389740.

[5]   Mitin, V.S.; Sharipov, E.I.; Mitin, A. v. High Deposition Rate Magnetrons: Key Elements and Advantages. **2013**, *22*, 5–10.

https://doi.org/10.1179/174329406X85038

[6]   Knotek, O.; Löffler, F.; Krämer, G. Process and Advantage of Multicomponent
       and Multilayer PVD Coatings. *Surface and Coatings Technology* **1993**, *59*, 14–20.
       https://doi.org/10.1016/0257-8972(93)90048-S

[7]   Şerban, V.A.; Roşu, R.A.; Bucur, A.I.; Pascu, D.R. Deposition of Titanium Nitride
       Layers by Electric Arc – Reactive Plasma Spraying Method. *Applied Surface
       Science* **2013**, *265*, 245–249. https://doi.org/10.1016/j.apsusc.2012.10.187

[8]   Urdas, V. Tratamente Termice, Termochimice, Coroziunea Metalelor Si Acoperiri
       de Suprafata; Univ: Sibiu, 2001; ISBN 9736512991.

[9]   Pawlak R; Korzeniewska E; Koneczny C; Hałgas B properties of thin metal layers
       deposited on textile composites by using the pvd method for textronic applications.
       https://doi.org/10.1515/aut-2017-0015

[10]   Korzeniewska, E.; de Mey, G.; Pawlak, R.; Stempień, Z. Analysis of Resistance to
       Bending of Metal Electroconductive Layers Deposited on Textile Composite
       Substrates in PVD Process. *Scientific Reports 2020 10:1* **2020**, *10*, 1–11.
       https://doi.org/10.1038/s41598-020-65316-2

[11]   Lee, S.J.; Huang, C.H.; Chen, Y.P. Investigation of PVD Coating on Corrosion
       Resistance of Metallic Bipolar Plates in PEM Fuel Cell. *Journal of Materials
       Processing Technology* **2003**, *140*, 688–693. https://doi.org/10.1016/S0924-
       0136(03)00743-X

[12]   Liu, C.; Leyland, A.; Bi, Q.; Matthews, A. Corrosion Resistance of Multi-Layered
       Plasma-Assisted Physical Vapour Deposition TiN and CrN Coatings. *Surface and
       Coatings Technology* **2001**, *141*, 164–173. https://doi.org/10.1016/S0257-
       8972(01)01267-1

[13]   Kim, H.; Jr., C.C.; Lavoie, C.; Rossnagel, S.M. Diffusion Barrier Properties of
       Transition Metal Thin Films Grown by Plasma-Enhanced Atomic-Layer
       Deposition. *Journal of Vacuum Science & Technology B: Microelectronics and
       Nanometer Structures Processing, Measurement, and Phenomena* **2002**, *20*, 1321.
       https://doi.org/10.1116/1.1486233

[14]   Navinšek, B.; Panjan, P.; Milošev, I. PVD Coatings as an Environmentally Clean
       Alternative to Electroplating and Electroless Processes. *Surface and Coatings
       Technology* **1999**, *116–119*, 476–487. https://doi.org/10.1016/S0257-
       8972(99)00145-0

[15]   Constantin, R.; Miremad, B. Performance of Hard Coatings, Made by Balanced

and Unbalanced Magnetron Sputtering, for Decorative Applications. *Surface and Coatings Technology* **1999**, *120–121*, 728–733. https://doi.org/10.1016/S0257-8972(99)00366-7

[16]   Beck, U.; Reiners, G.; Urban, I.; Jehn, H.A.; Kopacz, U.; Schack, H. Decorative Hard Coatings: New Layer Systems without Allergy Risk. *Surface and Coatings Technology* **1993**, *61*, 215–222. https://doi.org/10.1016/0257-8972(93)90228-G

[17]   Jehn, H.A. Improvement of the Corrosion Resistance of PVD Hard Coating–Substrate Systems. *Surface and Coatings Technology* **2000**, *125*, 212–217. https://doi.org/10.1016/S0257-8972(99)00551-4

[18]   Bi, X.; Lan, W.; Ou, S.; Gong, S.; Xu, H. Magnetic and Electrical Properties of FeSi/FeSi–ZrO2 Multilayers Prepared by EB-PVD. *Journal of Magnetism and Magnetic Materials* **2003**, *261*, 166–171. https://doi.org/10.1016/S0304-8853(02)01469-5

[19]   Åstrand, M.; Selinder, T.I.; Fietzke, F.; Klostermann, H. PVD-Al2O3-Coated Cemented Carbide Cutting Tools. *Surface and Coatings Technology* **2004**, *188–189*, 186–192. https://doi.org/10.1016/j.surfcoat.2004.08.021

[20]   Rossnagel, S.M. Sputter Deposition for Semiconductor Manufacturing. *IBM Journal of Research and Development* **1999**, *43*, 163–179. https://doi.org/10.1147/rd.431.0163

[21]   Selvakumar, N.; Barshilia, H.C. Review of Physical Vapor Deposited (PVD) Spectrally Selective Coatings for Mid- and High-Temperature Solar Thermal Applications. *Solar Energy Materials and Solar Cells* **2012**, *98*, 1–23. https://doi.org/10.1016/j.solmat.2011.10.028

[22]   Ko, R.K.; Park, C.; Kim, H.S.; Chung, J.K.; Ha, H.S.; Shi, D.; Song, K.J.; Yoo, S.I.; Moon, S.H.; Kim, Y.C. Fabrication of Meter-Long Coated Conductor Using RABiTS-PVD Methods. *IEEE Transactions on Applied Superconductivity* **2005**, *15*, 2707–2710. https://doi.org/10.1109/TASC.2005.847789

[23]   Pulker, H.K. Optical Coatings Deposited by Ion and Plasma PVD Processes. *Surface and Coatings Technology* **1999**, *112*, 250–256. https://doi.org/10.1016/S0257-8972(98)00764-6

[24]   Jacob, D.; Peiró, F.; Quesnel, E.; Ristau, D. Microstructure and Composition of MgF2 Optical Coatings Grown on Si Substrate by PVD and IBS Processes. *Thin Solid Films* **2000**, *360*, 133–138. https://doi.org/10.1016/S0040-6090(99)00738-5

[25]   Zhang, P.; Liu, J.; Xu, G.; Yi, X.; Chen, J.; Wu, Y. Anticorrosive Property of Al

Coatings on Sintered NdFeB Substrates via Plasma Assisted Physical Vapor
Deposition Method. *Surface and Coatings Technology* **2015**, *282*, 86–93.
https://doi.org/10.1016/j.surfcoat.2015.10.021

[26] Kaminski, J.; Tacikowski, M.; Brojanowska, A.; Kucharska, B.; Wierzchon, T.
The Effect of Tightening on the Corrosion Properties of the PVD Layers on
Magnesium AZ91D Alloy. *Journal of Surface Engineered Materials and
Advanced Technology* **2014**, *2014*, 270–281.
https://doi.org/10.4236/jsemat.2014.45031

[27] Lee, M.H.; Kim, Y.W.; Lim, K.M.; Lee, S.H.; Moon, K.M. Electrochemical
Evaluation of Zinc and Magnesium Alloy Coatings Deposited on
Electrogalvanized Steel by PVD. *Transactions of Nonferrous Metals Society of
China* **2013**, *23*, 876–880. https://doi.org/10.1016/S1003-6326(13)62542-X

[28] Cao, Y.; Zhang, P.; Sun, W.; Zhang, W.; Wei, H.; Wang, J.; Li, B.; Yi, X.; Xu, G.;
Wu, Y. Effects of Bias Voltage on Coating Structures and Anticorrosion
Performances of PA-PVD Al Coated NdFeB Magnets. *Journal of Rare Earths*
2021, *39*, 703–711. https://doi.org/10.1016/j.jre.2020.07.025

[29] Holleck, H.; Schier, V. Multilayer PVD Coatings for Wear Protection. *Surface and
Coatings Technology* **1995**, *76–77*, 328–336. https://doi.org/10.1016/0257-
8972(95)02555-3

[30] Lugscheider, E.; Bobzin, K. The Influence on Surface Free Energy of PVD-
Coatings. *Surface and Coatings Technology* **2001**, *142–144*, 755–760.
https://doi.org/10.1016/S0257-8972(01)01315-9

[31] Oettel, H.; Wiedemann, R. Residual Stresses in PVD Hard Coatings. *Surface and
Coatings Technology* **1995**, *76–77*, 265–273. https://doi.org/10.1016/0257-
8972(95)02581-2

[32] Bienk, E.J.; Reitz, H.; Mikkelsen, N.J. Wear and Friction Properties of Hard PVD
Coatings. *Surface and Coatings Technology* **1995**, *76–77*, 475–480.
https://doi.org/10.1016/0257-8972(95)02498-0

[33] Incerti, L.; Rota, A.; Valeri, S.; Miguel, A.; García, J.A.; Rodríguez, R.J.; Osés, J.
Nanostructured Self-Lubricating CrN-Ag Films Deposited by PVD Arc Discharge
and Magnetron Sputtering. *Vacuum* **2011**, *85*, 1108–1113.
https://doi.org/10.1016/j.vacuum.2011.01.022

[34] Fox-Rabinovich, G.S.; Yamomoto, K.; Veldhuis, S.C.; Kovalev, A.I.; Dosbaeva,
G.K. Tribological Adaptability of TiAlCrN PVD Coatings under High

Performance Dry Machining Conditions. *Surface and Coatings Technology* **2005**, *200*, 1804–1813. https://doi.org/10.1016/j.surfcoat.2005.08.057

[35]   Reiter, A.E.; Brunner, B.; Ante, M.; Rechberger, J. Investigation of Several PVD Coatings for Blind Hole Tapping in Austenitic Stainless Steel. *Surface and Coatings Technology* **2006**, *200*, 5532–5541. https://doi.org/10.1016/j.surfcoat.2005.07.100

[36]   Martín-Palma, R.J.; Lakhtakia, A. Vapor-Deposition Techniques. *Engineered Biomimicry* **2013**, 383–398. https://doi.org/10.1016/B978-0-12-415995-2.00015-5

[37]   Yan, H.; Hou, J.; Fu, Z.; Yang, B.; Yang, P.; Liu, K.; Wen, M.; Chen, Y.; Fu, S.; Li, F. Growth and Photocatalytic Properties of One-Dimensional ZnO Nanostructures Prepared by Thermal Evaporation. *Materials Research Bulletin* **2009**, *44*, 1954–1958. https://doi.org/10.1016/j.materresbull.2009.06.014

[38]   Nemetschek, R.; Prusseit, W.; Holzapfel, B.; Eickemeyer, J.; DeBoer, B.; Miller, U.; Maher, E. Continuous YBa2Cu3O7-Tape Deposition by Thermal Evaporation. *Physica C: Superconductivity* **2002**, *372–376*, 880–882. https://doi.org/10.1016/S0921-4534(02)00887-0

[39]   Jamkhande, P.G.; Ghule, N.W.; Bamer, A.H.; Kalaskar, M.G. Metal Nanoparticles Synthesis: An Overview on Methods of Preparation, Advantages and Disadvantages, and Applications. *Journal of Drug Delivery Science and Technology* **2019**, *53*, 101174. https://doi.org/10.1016/j.jddst.2019.101174

[40]   Zhu, Y.; Mendelsberg, R.J.; Zhu, J.; Han, J.; Anders, A. Transparent and Conductive Indium Doped Cadmium Oxide Thin Films Prepared by Pulsed Filtered Cathodic Arc Deposition. *Applied Surface Science* **2013**, *265*, 738–744. https://doi.org/10.1016/j.apsusc.2012.11.096

[41]   Sanders, D.M.; Anders, A. Review of Cathodic Arc Deposition Technology at the Start of the New Millennium. *Surface and Coatings Technology* **2000**, *133–134*, 78–90. https://doi.org/10.1016/S0257-8972(00)00879-3

[42]   Tay, B.K.; Zhao, Z.W.; Chua, D.H.C. Review of Metal Oxide Films Deposited by Filtered Cathodic Vacuum Arc Technique. *Materials Science and Engineering: R: Reports* **2006**, *52*, 1–48. https://doi.org/10.1016/j.mser.2006.04.003

[43]   Baptista, A.; Silva, F.; Porteiro, J.; Míguez, J.; Pinto, G. Sputtering Physical Vapour Deposition (PVD) Coatings: A Critical Review on Process Improvement and Market Trend Demands. *Coatings 2018, Vol. 8, Page 402* **2018**, *8*, 402. https://doi.org/10.3390/coatings8110402

[44]   Mattox, D.M. Ion Plating — Past, Present and Future. *Surface and Coatings Technology* **2000**, *133–134*, 517–521. https://doi.org/10.1016/S0257-8972(00)00922-1

[45]   Ion Plating. *Handbook of Deposition Technologies for Films and Coatings* **2010**, 297–313. https://doi.org/10.1016/B978-0-8155-2031-3.00006-5

[46]   Tai, C.N.; Koh, E.S.; Akari, K. Macroparticles on TiN Films Prepared by the Arc Ion Plating Process. *Surface and Coatings Technology* **1990**, *43–44*, 324–335. https://doi.org/10.1016/0257-8972(90)90085-Q

[47]   Mattox, D.M. Fundamentals of Ion Plating. *Journal of Vacuum Science and Technology* **2000**, *10*, 47. https://doi.org/10.1116/1.1318041

[48]   Matsue, T.; Hanabusa, T.; Ikeuchi, Y. The Structure of TiN Films Deposited by Arc Ion Plating. *Vacuum* **2002**, *66*, 435–439. https://doi.org/10.1016/S0042-207X(02)00167-7

[49]   Mattevi, C.; Kim, H.; Chhowalla, M. A Review of Chemical Vapour Deposition of Graphene on Copper. *Journal of Materials Chemistry* **2011**, *21*, 3324–3334. https://doi.org/10.1039/C0JM02126A

[50]   Bryant, W.A. The Fundamentals of Chemical Vapour Deposition. *Journal of Materials Science 1977 12:7* **1977**, *12*, 1285–1306. https://doi.org/10.1007/BF00540843

[51]   Choy, K.L. Chemical Vapour Deposition of Coatings. *Progress in Materials Science* **2003**, *48*, 57–170. https://doi.org/10.1016/S0079-6425(01)00009-3

[52]   Dobrzański, L.A.; Lukaszkowicz, K.; Paku³a, D.; Miku³a, J. Archives of Materials Science and Engineering Corrosion Resistance of Multilayer and Gradient Coatings Deposited by PVD and CVD Techniques. **2007**.

[53]   Dobrzański, L.A.; Lukaszkowicz, K.; Paku³a, D.; Miku³a, J. Archives of Materials Science and Engineering Corrosion Resistance of Multilayer and Gradient Coatings Deposited by PVD and CVD Techniques. **2007**.

[54]   Baptista, A.; Silva, F.; Porteiro, J.; Míguez, J.; Pinto, G. Sputtering Physical Vapour Deposition (PVD) Coatings: A Critical Review on Process Improvement and Market Trend Demands. *Coatings 2018, Vol. 8, Page 402* **2018**, *8*, 402. https://doi.org/10.3390/coatings8110402

[55]   Dobrzanski, L.A.; Lukaszkowicz, K.; Mikuła, J.; Pakuła, D. Structure and Corrosion Resistance of Gradient and Multilayer Coatings. *Journal of Achievements in Materials and Manufacturing Engineering* **2006**, *18*, 75–78.

[56] Elsener, B.; Rota, A.; Böhni, H. Impedance Study on the Corrosion of PVD and CVD Titanium Nitride Coatings. *Materials Science Forum* **1989**, *44–45*, 29–38. https://doi.org/10.4028/www.scientific.net/MSF.44-45.29

[57] Kimapong, K.; Poonayom, P.; Wattanajitsiri, V. Microstructure and Wear Resistance of Hardfacing Weld Metal on JIS-S50C Carbon Steel in Agricultural Machine Parts. In Proceedings of the Materials Science Forum; Trans Tech Publications Ltd, 2016; Vol. 872, pp. 55–61. https://doi.org/10.4028/www.scientific.net/MSF.872.55

[58] Özkan, D.; Kaleli, H. Surface and Wear Analysis of Zinc Phosphate Coated Engine Oil Ring and Cylinder Liner Tested with Commercial Lubricant. *Advances in Mechanical Engineering* **2014**, *2014*. https://doi.org/10.1155/2014/150968

[59] Wang, Q.; Luo, S.; Wang, S.; Wang, H.; Ramachandran, C.S. Wear, Erosion and Corrosion Resistance of HVOF-Sprayed WC and Cr3C2 Based Coatings for Electrolytic Hard Chrome Replacement. *International Journal of Refractory Metals and Hard Materials* **2019**, *81*, 242–252. https://doi.org/10.1016/j.ijrmhm.2019.03.010

[60] Cunha, L.; Moura, C.; Vaz, F.; Moura, C.; Vaz, F. Functional Coatings-a Look into the State of the Art of Hard and Decorative Coatings Development of New Hard Decorative Coatings Based on Transition Metal Oxynitrides View Project Functional Coatings-a Look into the State of the Art of Hard and Decorative Coatings. **2003**.

[61] Tański, T. Surface Layers on the Mg-Al-Zn Alloys Coated Using the CVD and PVD Methods Manufacturing and Processing.

About the Authors

Diana Petronela BURDUHOS-NERGIS

Assistant Professor PhD.Eng.
Department of Materials Engineering and Industrial Safety, Faculty of Materials Science and Engineering, "Gheorghe Asachi" Technical University of Iasi
diana.burduhos@tuiasi.ro, www.afir.org.ro/dpbn

Researcher and assistant professor at Gheorghe Asachi Technical University of Iasi, Faculty of Materials Science and Engineering, with a doctoral thesis on the study and improvement of carbon steel components in personal protective equipment by depositing different types of coatings, being involved in scientific research since she was a student. She has over 18 publication, 14 of them indexed by Web of Science. She has many awards received from presentations at conferences or invention exhibitions.

Dumitru-Doru BURDUHOS-NERGIS

Assistant Professor PhD.Eng.
Department of Materials Engineering and Industrial Safety, Faculty of Materials Science and Engineering, "Gheorghe Asachi" Technical University of Iasi
doru.burduhos@tuiasi.ro, https://www.afir.org.ro/ddbn/

Materials engineering researcher with 5 years of experience in the field of geopolymers. The research activity in the field carried out during the elaboration of the thesis for the master's degree graduation, was continued within the PhD stage, starting with 2017, and the scientific research results were disseminated in a number of 24 publications, of which 7 articles were published in Web of Science (WoS) (3 as the first author), 4 articles published in Proceedings, indexed WOS, 6 in Conference Proceedings indexed SCOPUS, 5 articles published in BDI-listed journals, one international Book and one international book chapter.

Simona-Madalina BALTATU

Lecturer Ph.D. Eng.
Department of Technology and Equipment for Materials Processing, Faculty of Materials Science and Engineering, "Gheorghe Asachi" Technical University of Iasi
cercel.msimona@yahoo.com, http://www.afir.org.ro/msb/

PhD Eng. from 2014 she focused on developing new biomaterials and advanced characterization. She published over 30 articles, 2 books, one international book chapter, 5 patent applications and she carry out activities in 6 projects.

Petrica VIZUREANU

Professor Ph.D. Eng.

Head of department at Department of Technology and Equipment for Materials Processing, Faculty of Materials Science and Engineering, "Gheorghe Asachi" Technical University of Iasi

peviz2002@yahoo.com, http://afir.org.ro/peviz/

Professor and researcher at "Gheorghe Asachi" Technical University of Iasi, with more than 30 years of experience. Ph.D. degree, since 1999 in Materials science and engineering; 2010 - present Ph.D. Supervisor in Materials Engineering domain. He has over 150 publications, 130 articles being indexed in ISI Web of Science. He has large experience in the field of composite materials; ceramic materials, insulating materials; optimization of materials characteristics. H-index is 16.

www.ingramcontent.com/pod-product-compliance
Lightning Source LLC
Chambersburg PA
CBHW071657210326
41597CB00017B/2230